Saskia Reibe

Development and Identification of Corpse-Associated Calliphoridae

Saskia Reibe

Development and Identification of Corpse-Associated Calliphoridae

Consequences for Estimating the Post-Mortem Interval in Medico-Legal Investigations

Südwestdeutscher Verlag für Hochschulschriften

Impressum/Imprint (nur für Deutschland/ only for Germany)
Bibliografische Information der Deutschen Nationalbibliothek: Die Deutsche Nationalbibliothek
verzeichnet diese Publikation in der Deutschen Nationalbibliografie; detaillierte bibliografische
Daten sind im Internet über http://dnb.d-nb.de abrufbar.
Alle in diesem Buch genannten Marken und Produktnamen unterliegen warenzeichen-, marken-
oder patentrechtlichem Schutz bzw. sind Warenzeichen oder eingetragene Warenzeichen der
jeweiligen Inhaber. Die Wiedergabe von Marken, Produktnamen, Gebrauchsnamen,
Handelsnamen, Warenbezeichnungen u.s.w. in diesem Werk berechtigt auch ohne besondere
Kennzeichnung nicht zu der Annahme, dass solche Namen im Sinne der Warenzeichen- und
Markenschutzgesetzgebung als frei zu betrachten wären und daher von jedermann benutzt
werden dürften.

Verlag: Südwestdeutscher Verlag für Hochschulschriften GmbH & Co. KG
Dudweiler Landstr. 99, 66123 Saarbrücken, Deutschland
Telefon +49 681 37 20 271-1, Telefax +49 681 37 20 271-0
Email: info@svh-verlag.de
Zugl.: Bonn, Rheinische Friedrich-Wilhelms Univerität, Diss., 2010

Herstellung in Deutschland:
Schaltungsdienst Lange o.H.G., Berlin
Books on Demand GmbH, Norderstedt
Reha GmbH, Saarbrücken
Amazon Distribution GmbH, Leipzig
ISBN: 978-3-8381-2231-1

Imprint (only for USA, GB)
Bibliographic information published by the Deutsche Nationalbibliothek: The Deutsche
Nationalbibliothek lists this publication in the Deutsche Nationalbibliografie; detailed
bibliographic data are available in the Internet at http://dnb.d-nb.de.
Any brand names and product names mentioned in this book are subject to trademark, brand
or patent protection and are trademarks or registered trademarks of their respective holders.
The use of brand names, product names, common names, trade names, product descriptions
etc. even without a particular marking in this works is in no way to be construed to mean that
such names may be regarded as unrestricted in respect of trademark and brand protection
legislation and could thus be used by anyone.

Publisher: Südwestdeutscher Verlag für Hochschulschriften GmbH & Co. KG
Dudweiler Landstr. 99, 66123 Saarbrücken, Germany
Phone +49 681 37 20 271-1, Fax +49 681 37 20 271-0
Email: info@svh-verlag.de

Printed in the U.S.A.
Printed in the U.K. by (see last page)
ISBN: 978-3-8381-2231-1

Preface

This book is a reprint of my PhD-Thesis I finished in June 2010. Therefore, the bibliography does not contain any literature more recent. Also, the experiments were designed to improve forensic entomology in Germany. The flies used for molecular species identification and growth modelling were captured and bred in Bonn, Germany. Hence, the developmental data for blow flies shown here should be used with reservation in other countries.

I personally think that forensic entomology is a field worth the effort as it combines classical biological research and forensic thinking. The field is highly underrepresented in universities and the juridical system not only but especially in Germany. Therefore, I am highly delighted about every reader of this book hoping the field will evolve over time.

Cologne, December 31st, 2010

Contents

IV Calculating larval age to estimate a post-mortem interval

6 A new simulation-based model for calculating post-mortem intervals using developmental data for *Lucilia sericata* (Dipt.: Calliphoridae)

7 Growth modeling of *Calliphora vicina*

Part I

Introduction

Chapter 1

Introduction

1.1 Forensic science - a survey

Forensic science is no subject included in any german university curriculum. Nonetheless, it is applied in every homicide investigation in the country - not only by the police men but mainly by scientists using methods from natural sciences. Among them, biologists analyzing DNA material from crime scenes or toxicologists examining drug samples. To analyze complex blood spatter patterns, geometrical and mathematical laws are applied. The used methods were not established to investigate crime cases in the first place but were rather adapted to use them as such.

Every crime case is an individual case. This is one of the most important flaws but also challenges in real crime cases: repetition is possible only rarely and double-blinded studies are impossible. However, methods and experimental designs originating from natural sciences can very well be repeated and tested double-blinded. In court, objective evidence is much more significant than witness statements as in most cases reliabilities of experimental results can be calculated. In general, forensic science uses the natural sciences to be applied in investigating crime cases. This is also a fact for the field of entomology. As a university subject classic entomology was gradually replaced by modern molecular aspects as DNA barcoding (Hajibabaei et al., 2007).

However, for forensic investigations entomological knowledge can be of value as the identification and assignment of arthropods found in relation with crime scenes and corpses can help gaining information for instance on a minimum post-mortem interval (PMI) (Catts and Goff, 1992). The necessary data has been produced in the

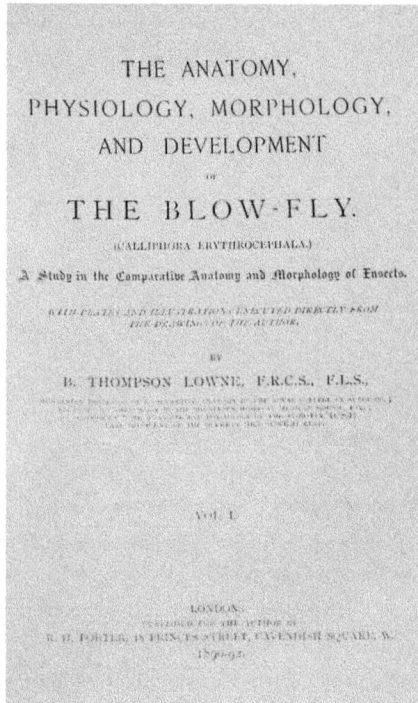

Figure 1.1: Titlepage: Lowne (1890) The anatomy, physiology, morphology and development of the blow-fly.

last decades by scientists not knowing that their results will become helpful in crime scene investigations. Their research was directed towards a general understanding of physiology, ecology, behavior and development of arthropods (Fig. 1.1) (Beattie, 1928; Feist, 1926; Janisch, 1928, 1931; Lowne, 1890; Mellanby, 1938; Smirnov and Zhelochovtsev, 1927; Weinland, 1906).

In the field of forensic entomology the exact opposite happens to what is known from molecular sciences: old knowledge has to be reactivated, as it is still valuable, to avoid wasting time by duplicating research. On the one hand, several old publications can be found and information can be gained, on the other hand a lot of new research is additionally necessary to get forensic entomology started. It is important to transfer the knowledge based on experiments with animal carcasses (Lane, 1975;

Putmann, 1977; Steinborn, 1981) to real cases with human bodies and to work on the reliability of statements based on insect evidence that are used in court eventually. A lot of effort has to be put in statistical analysis and also in research about local insect fauna. In Germany, the first case in court involving and partially basing a verdict on insect evidence was the case of Klaus Geyer in 1997 (Benecke and Seifert, 1999). K. Geyer was convicted after a combined analysis of the maggots collected from the corpse of his murdered wife to estimate the post-mortem interval as a time point where the accused had no alibi and a person-location match due to a specimen of an ant adhering to his rubber boots. This particular species could only originate from a nest nearby where the corpse was found. Since then, only a few scientists tried to establish the field of forensic entomology in Germany either in institutes for forensic medicine or as freelancers. However, forensic entomology is far from being implemented in the system as a frequently requested method. Nevertheless, a lot of interesting research needs to be done to improve the field not only in Germany but worldwide. Especially, the reliabilities of forensic entomological methods have to be determined to strengthen the status of the field in court. In a case from 1965 from England, the medico-legal doctor determined a PMI based on insect evidence but three different witnesses saw the dead person after the supposedly time of death. The court believed the medico-legal doctor and convicted the accused for murder. The medico-legal doctor stated afterwards, that this trial was pretty satisfying to him and as he was well known for estimating PMIs, it would have been a public disgrace for him, if he had been wrong (Simpson, 1986). This case was reviewed by Henssge and Madea (1988) in an article on subjective and objective problems in estimating time since death. These authors stated that quality control for the correctness of an expert witness statement can not be the success of a trial. Consequently, the methods an expert witness statement is based on must be validated and reliable.

1.2 Forensic entomology 25 years ago and today - any improvements?

Historical events in forensic entomology reaching back to the 13th century have been described extensively in several studies, publications and reviews and will not be repeated here (Amendt et al., 2000, 2004; Benecke, 2001, 2005, 2008). To cover recent historical events in the establishment of forensic entomology as a specific field, a review from 1985 written by Keh is used as starting point for the retrospective (Keh, 1985). In general, after the review of Keh about 500 publications related to the term "forensic entomology" were published until today (authors bibliography, in PubMed about 260 are listed: http://www.ncbi.nlm.nih.gov). In comparison, searching the term "diabetes/insulin" which is a highly relevant search term for health- and aging research, reveals about 500 papers published alone between October and November 2009 (http://www.ncbi.nlm.nih.gov). This shows how small the research field forensic entomology is in comparison.

Keh states that "the term 'forensic entomology,' though not strictly defined, is generally applied to the study of insects and other arthropods associated with certain suspected criminal events, for the purpose of uncovering information useful to an investigation.' This description is accurate. Keh reports that forensic entomology was introduced as a corrobation to methods used by forensic pathologists for approximating the time of recent death, as their methods such as body temperature, livores, rigor mortis and stages of decomposition are limited in their power.

He claims that for most experiments animal carcasses are used and that results might have to be adapted to real cases. This is a fact till today: apart from experiments at the University of Tennessee Anthropological Research Facility where human corpses are exposed to decompose the usual way is to use animal carcasses. On the so called body farm insect succession is not the primary aim of investigation but guest researchers are sometimes accepted (Schoenly et al., 2007; Shahid et al., 2003).

Keh addresses the fact that "insects are often reared in laboratories under constant temperature and humidity to determine time required for their development, but in nature fluctuating temperatures, which may hasten, retard, or have no effect on speed of development, are more often encountered". In 2006, Donovan et al. published their work about larval growth rates of *C. vicina* over a range of temperatures (Donovan et al., 2006), using constant temperatures. They discussed different studies from other authors and presented their data concerning fluctuating temperatures. Eventually, they concluded that fluctuating temperatures can affect growth in one way or another. However, no solution was presented how to handle the problem. Till now, the biggest problem in applying growth data in actual crime cases results from the fact that not enough data exists. For species originating from Germany no data exists at all, let alone data produced under fluctuating temperatures even from other countries.

Furthermore, Keh mentions the aspect of larval aggregations and the following heating up in such masses, that may influence the growth of the larvae, as they develop poikilothermically. The most recent work addressing larval masses on carcass is published by Slone and Gruner in 2007. They correlated the internal temperature of a mass to its volume and compared it with the ambient temperature. They showed that the larger the volume, the higher the internal temperature up to temperatures lethal for the larvae. Whereas larvae from the center of large masses move to the cooler periphery and also evaporative cooling of the wet larvae would occur, therefore the average temperature experienced by a larva would be less than the hottest temperatures in the aggregations. However, they did not show in what way their results would improve the existing methods to estimate a post-mortem interval.

Finally, Keh wrote "bodies are found indoors as well as outdoors, in various types of containers, wrapped in a variety of materials, etc."..."since critical elements are either unexplored or unreported". Apart from some experiments comparing animal carcass exposed in a vehicle with carcass exposed outdoors (Voss et al., 2008) this thesis addresses the experimental question of indoor insect colonization of carcass for the first time (Reibe and Madea, 2010a). The first experiments comparing rotten meat lying open to meat covered with gauze were performed by Frencesco Redi in 1668. He wanted to prove that maggots can not be generated in the rotten meat spontaneously, so he observed rotten meat placed in jars either open or covered. The meat in the covered jars was not infested with maggots in contrast to the meat in the open jars (Redi, 1674).

In 1992 a second review on forensic entomology was published (Catts and Goff, 1992). At that time, two books dealt specifically with forensic entomology (Haskell and Catts, 1990; Smith, 1986). The book published by Smith is till today one of the most important books on the topic including an identification key to several arthropod families associated with carcass. Catts states that several case reports were published since 1985, showing the applicability of the field. Especially the work of P. Nuorteva and M. Leclercq brought progress in Europe by combining entomology with medico-legal cases (Leclercq, 1983, 1988a,b,c; Leclercq and Brahy, 1990; Leclercq et al., 1991; Nuorteva, 1965; Nuorteva et al., 1974).

Furthermore, a basic principle of PMI estimation is mentioned by Catts: summing of heat values as accumulated degree hours (ADH). In that context he stresses that 'the need for very careful experimental design and data recording in establishing baseline data on fly development cannot be overemphasized. Eventually computer modeling of maggot development may refine the accuracy of estimating maggot age'. Today, these statements are as relevant as 18 years ago. Catts also reports that since 1985 more experiments e.g. on succession data have been performed to understand the behavior of certain arthropod families towards different decomposition stages. Nevertheless, they are only of value when intraspecific variations or microclimatic differences are considered to assure a comparability for other cases.

Catts mentions that 'recently, attention has focused on the age determination of puparia, as Greenberg points out that as much as 40% of the blow fly life span is spent as puparium.' This problem is still unsolved. A study group from Frankfurt/Main addresses the problem by applying molecular methods such as fishing for developmental genes that are switched on or off during pupal stage (Boehme et al., 2009). In general, molecular approaches such as species identification are of growing interest in the field. Catts reports 'recent advances that allow DNA probe identification of insects or their isolated body fragments, either fresh or dried, might be applicable to discriminating among species of maggots where morphological distinctions are lacking'. This prediction became true. Several publications deal with molecular species identification mostly of blow flies originating from all over the world (Benecke, 1998; Chen et al., 2004; Desmyter and Gosselin, 2009; Harvey et al., 2008, 2003; Malgorn and Coquoz, 1999; Sperling et al., 1994; Wells and Stevens, 2008). However, prior to this thesis no results were published concerning molecular species identification of german blow flies.

To sum up, although some ground was covered in the field of forensic entomology in the last 25 years, a lot of work still needs to be done. Moreover, even if the field is considered to be global, more scientists in each geographical region should work on producing region-specific data that can be used in criminal investigations.

1.2.1 Trends in forensic entomology

A lot of research has to be done in the establishment of growth data, certain behavioral aspects of the corpse-associated arthropods and practical aspects of PMI estimation. Nevertheless, judging from recent publications the trends point to the extension of the field towards other methods and arthropods. A new aspect is the investigation of mites associated with carcass (Perotti and Braig, 2009). Also a new field of investigation is the analysis of the smell related to certain stages of decay and which components act as attractants for the arthropods (Dekeirsschieter et al., 2009; Kalinová et al., 2009). Furthermore, the repelling effect of some household products on fly attraction to cadavers is of interest (Charabidze et al., 2009). An improvement of PMI estimation should be made by analyzing hydrocarbons from the cuticle of different stages of blow flies (Moore, 2009; Roux et al., 2008). At last, other methods of identification are tested in extension to the common molecular methods as analysis of the COI region: antigen-based rapid diagnostic test (McDonagh et al., 2009).

1.3 Basics of forensic entomology

1.3.1 Succession

All decaying organic material, including corpses, is a natural habitat for several arthropods. They use it as nourishment, breeding site, mating or hiding place. As the material decomposes, it undergoes a series of changes offering different species exactly what they are specialised on. Several succession studies were carried out in several different countries (Anderson and VanLaerhoven, 1996; Archer, 2003; Archer and Elgar, 2003; Arnaldos et al., 2001; Bharti and Singh, 2003; Eberhardt and Elliot, 2008; Grassberger and Frank, 2004; Watson and Carlton, 2005) to understand the order in which species response to the different stages of decomposition and to further correlate species and decomposition stage to estimate a post-mortem interval in real cases (Goff, 1993). For the studies, animal carcasses were exposed to monitor the different stages of decay and to sample the corresponding arthropods. Nevertheless, all series have to be regarded as individual as the habitat and the microclimate can have serious effects on the insect fauna.

Figure 1.2 shows exemplarily different stages of decay of human corpses. All of the corpses were autopsied in the Institute of Forensic Medicine, Bonn, Germany: 1. fresh (a few hours after death), 2. active decay (2 weeks after death), 3. advanced decay (3 weeks after death), 4. advanced decay (2 months after death). 5. dry or skeletonized (2 years after death). All corpses have been found outdoors but under different microclimatic influences, however, the general progress of decay can

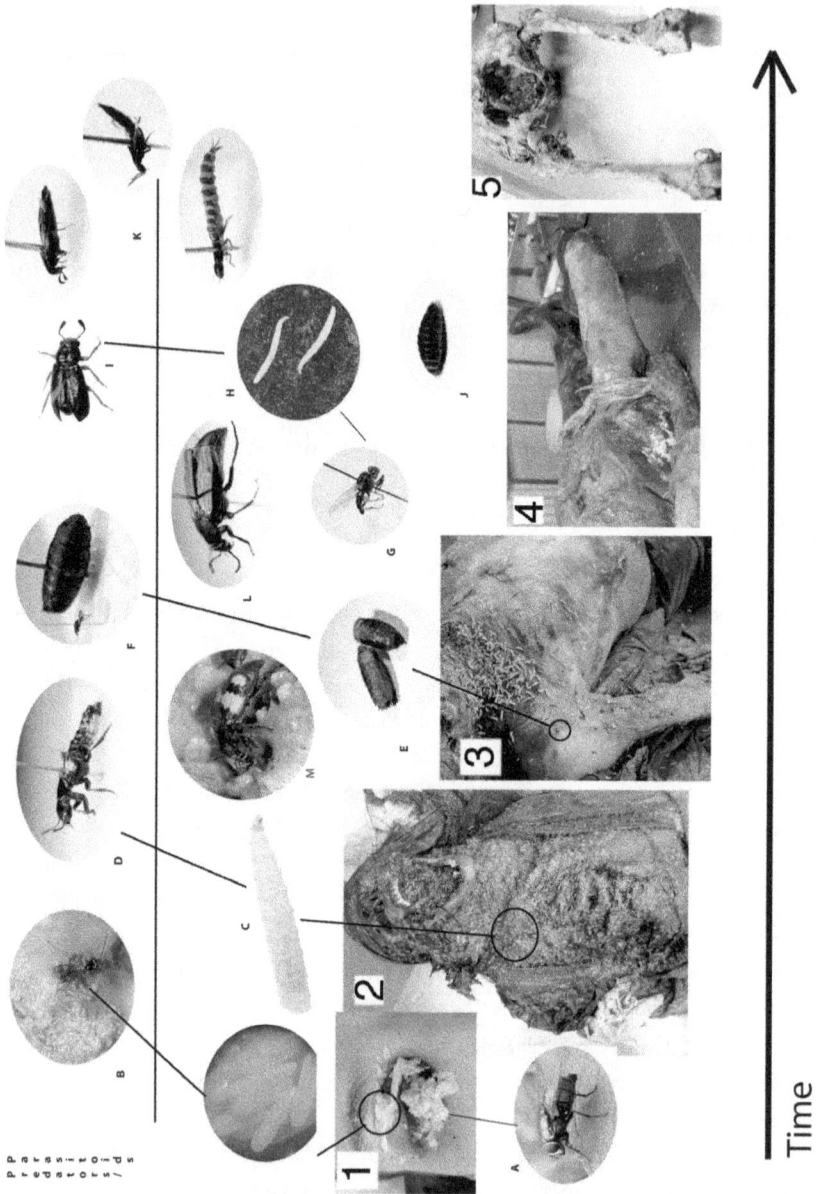

Figure 1.2: Visitation of several interacting insect groups in different stages of decomposition. More explanation in the text.

be followed. In figure 1.2, specimens of the visiting species are grouped beneath the stage of decomposition in which they were observed.

Among the earliest visitors of a corpse are the blow flies (Dipera: Calliphoridae) (A). They deposit their eggs in areas where the hatching first instar larvae can feed (Norris, 1965). Another early visitor is *Alysia manducator* (Hymenoptera: Braconidae) (B). It is attracted to the odor of decaying meat and parasites the larvae of blow flies (Matthews, 1974). The latter feed on the decomposing tissue which is softened and cracked down by the activity of bacteria. A predator of the feeding larvae is *Creophilus maxilosus* (Coleoptera: Staphilinidae) (D) (Greene, 1996). The next developmental step for the blow fly larvae is the pupariation; in the pupal stage (E) the metamorphosis form larvae to adult blow fly takes place (Denlinger, 1994). A common parasitoid of the puparium is *Nasonia vitripennis* (Hymenoptera: Pteromalidae) (F) (Fabritius and Klunker, 1991). After parasitation, one puparium can be nutrition site for hundreds of new specimens of *Nasonia vitripennis*, as the wasps are very small (see size comparison in Fig. 1.2 F). In an advanced stage of decay, Piophilidae (Diptera) are attracted to the corpse (G). Their larvae (H) have the ability to jump, as a defensive mechanism against their predators *Necrobia rufipes* (Coleoptera: Cleridae) (I). In general, the longer the decomposition proceeds the more arthropod families arrive to benefit from the time-dependent habitat. Stage 4 also attracted several species from the family of Staphilinids (K). They all prey on dipterous larvae. Another beetle is frequently found in later stages of decay: *Nicrodes littoralis* (Coleoptera: Silphidae) (L). Its larvae can be found on cadavers a few weeks after death (Matuszewski et al., 2008). Addtionally, other dipteran families can be observed e.g. Muscidae. In stage 4, pupae of *Fannia manicata* (Diptera: Fanniidae) were found (J). The larvae also feed on decaying material.

The last species shown in Figure 1.2 is *Necrophorus vespilloides* (Coleoptera: Silphidae) (M). The picture was taken during an experiment exposing small piglet carcasses. The beetle is associated with carcass but is known for burying small vertebrate carcasses and preparing them as food for their young (Kalinová et al., 2009; Scott, 1998). Therefore, they are mostly observed in succession experiments with small animal carcasses. Human cadavers are too big, so that *N. vespilloides* has not been reported in real forensic cases. This is one of the problems when succession studies are carried out with small animal carcasses, it is only partially representative. Also, *Lucilia ampullacae* was frequently caught on piglet carcasses exposed in Bonn

11

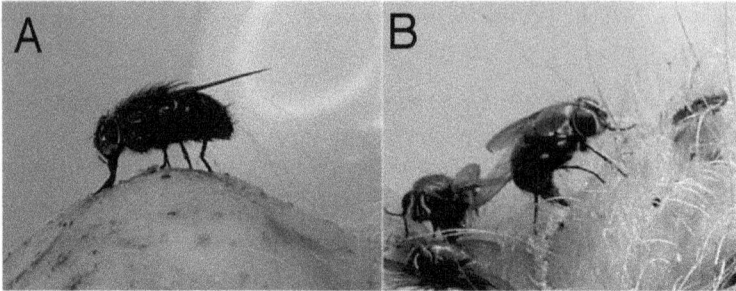

Figure 1.3: Blow flies on carcass: A Extended proboscis of blow fly. B Extended ovipositor of blow fly.

but never in a real case. Furthermore, succession studies are conducted in several different places all around the planet which makes it difficult to compare them and to use them in other parts. To understand the local fauna it is strongly recommended to conduct such experiments and to compare the results to the insects actually found on corpses investigated in the local institute for forensic medicine.

Another crucial point is monitoring species that occur rather in indoor locations. Specimens of *Megaselia scalaris* (Diptera: Phoridae) are most often found on corpses lying indoors, as they are very small and can enter even enclosed environments (Manlove and Disney, 2008). Also *Fannia cannicularis* is often found indoors, its trivial name is *little housefly. Fannia scalaris* occurs indoors whenever the circumstances are primitive and neglected and as it is associated with lavatories and cesspits, it is also called *latrine fly* (Benecke and Lessig, 2001).

To sum up, succession studies can be a useful tool to classify species in their relation to certain decomposition stages although no clear cuts can be made from one stage to another. Moreover, every succession study is highly dependent on the habitat and microclimate. Nevertheless, it is important to know the local fauna to be able to judge the entomological findings in each individual case.

1.3.2 Blow flies and their behavior towards carcass

Blow flies (Calliphoridae) are attracted to carcass as decomposing material is a protein source for themselves as well as breeding site for their progeny (Lane, 1975; Putmann, 1977; Smith and Wall, 1997). If a freshly dead piglet is exposed in the

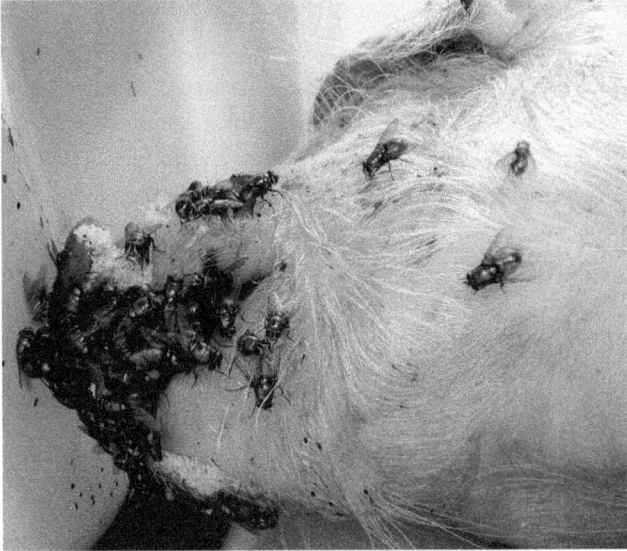

Figure 1.4: Blow flies forming aggregations on carcass to oviposit.

open field the first blow flies arrive within minutes. Two different behaviors of blow flies can be observed: extending their proboscis to absorb liquid (Fig. 1.3 A) or extending their ovipositor at distinct regions to find a proper place to oviposit (Fig. 1.3 B). If the pregnant blow fly found a suitable place it deposits their eggs in a clutch. After a few hours it can be observed that blow flies prefer spots where eggs have already been deposited to oviposit as well so that big egg aggregations are formed (Browne et al., 1969). Also it seems that blow flies accumulate at convenient spots on the carcass (Fig. 1.4); accumulated flies act as an attractant for other flies (Norris, 1965). In general, blow flies try to find oviposition sites that offer the best conditions for their progeny. As the hatching first instar larvae can only feed on moist tissue, the blow flies choose the body's natural orifices or wounds to oviposit.

To improve PMI determination using developmental data of blow fly larvae (see 1.5) it is important to estimate the time of oviposition correctly. At nighttime blow flies under most circumstances will not be active to find carcass and to oviposit (Amendt et al., 2008; Greenberg, 1990; Singh and Bharti, 2001; Wooldridge et al., 2007). Furthermore, if a corpse is stored in an enclosed environment, the blow flies

13

will reach it with a delay of 24 hours when a window is tilted or even later when windows and doors are closed (Reibe and Madea, 2010a; Reibe et al., 2008).

There have been approaches to characterize preferences of the different blow fly species toward ecological circumstances such as photoperiod and temperature (Vinogradova and Kaufman, 1995), sun or shade (Joy et al., 2006; Sharanowski et al., 2008; Shean et al., 1993) and habitat (Grassberger and Frank, 2004). By using meat traps in different habitats and monitoring the captured species their ecological preferences were described (Nuorteva, 1967; Steinborn, 1976; Steiner, 1948). Nevertheless, the results can only be used with caution as they can vary in each geographical region.

Furthermore, the larvae of the different species can also exhibit distinct behavior. The larvae of *Protophormia terraenovae* for example pupate on the surface of the feeding substrate (Grassberger and Reiter, 2002) whereas other species leave the feeding substrate and bury themselves in the ground or hide to pupate. All blow fly larvae are typically photonegative and will avoid light by crawling away.

1.3.3 Identification of insects

Most important in forensic case work involving entomological evidence is the identification of the insect species collected in association with the corpse or the surroundings. Identification is the foundation of all further insect based estimations. One quick and easy way is the morphological identification using appropriate identification keys. The standard reference for the identification of blow flies is the book *Blowflies (Diptera, Calliphoridae) of Fennoscandia and Denmark* by K. Rognes from 1991. Another standard reference is *A Manual of Forensic Entomology* by K.G.V. Smith, it includes an identification key for adults and larvae associated with carrion. The only equipment necessary for morphological species determination is a dissecting microscope with a proper light source.

However, it may be impossible to identify an insect by means of its morphology e.g. due to damage, then it might be possible to use molecular identification tools. To ensure correct species identification, established molecular methods were transferred to the forensic field (Benecke, 1998; Sperling et al., 1994; Stevens and Wall, 1996, 1997; Wallman and Adams, 1997). Analysis of mitochondrial DNA (mtDNA) and particulary of the cytochrome oxidase I gene (COI) appears to be a useful tool in species identification among the subfamilies of Calliphoridae (Harvey et al., 2008, 2003; Wallman et al., 2005; Wells and Williams, 2005; Wells et al., 2007). Although the method is reliable, no species originating from Germany were tested and their sequences compared to those from other countries. The method was tested for German blow flies for the first time during this thesis (Reibe et al., 2009).

1.4 Physiology of larval development

1.4.1 Temperature

Insects are poikilothermic organisms. Their body temperature changes with ambient temperature. The metabolic rate of poikilotherm animals shows the same temperature dependency as the reaction kinetics in a biochemical system: the rate of a chemical reaction is increased twofold for each rise of 10 °C in temperature (van't Hoff's reaction-rate-temperature rule) (Wehner and Gehring, 1995). For insect development a relationship between ambient temperature and duration of developmental processes is long known. There is a temperature zone where the development rate is optimal, furthermore, temperature thresholds exist below or above that optimum where no development will take place. During the last century a lof of effort has been made to describe larval growth correctly and to find a function representing the temperature dependent growth rate best (Janisch, 1928). Several empirical and physiochemical formulations of development were proposed and two were discussed in particular by Sharpe and De Michele in 1977 to present the reaction kinetics of poikilotherm development: the day-degree or temperature summation rule and the non-linear temperature inhibition model (Sharpe and DeMichele, 1977). The day-degree summation rule assumes that the rate of development is proportional to temperature:

$$k = b(T - T_0) \tag{1.1}$$

where k is the rate of development, b is a constant, T the ambient temperature and T_0 is a species-specific value, the so called development zero, which is the x-intercept, i.e., an extrapolation of the linear approximation of the reciprocal of time for development. Figure 1.5 shows the developmental times for the egg stage of D. *melanogaster* (white circles): the higher the temperature the shorter the developmental time in hours. Also shown is the developmental rate [1/dev time], a S-shaped curve, as well as the approximately linear region of the developmental curve where

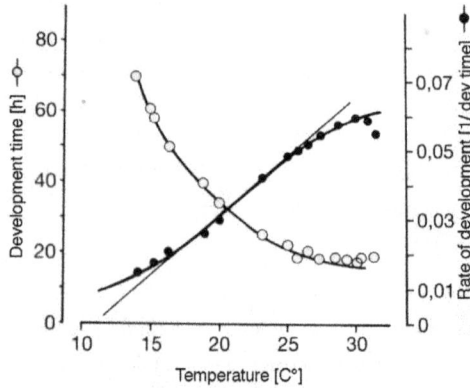

Figure 1.5: Duration in hours of the egg development of *D. melanogaster* for different temperatures and the corresponding developmental rate [1/dev time], adapted from Hoffmann (1995).

the day-degree concept is valid. In this concept T_0 is determined by prolonging the linear portion of the curve till it crosses the x-axis (for example in Fig. 1.5 about 12 °C). Sharpe and DeMichele state that a) T_0 calculated like this is not the true threshold for development, it has to be determined experimentally and b) severe errors in predicted development rates occur at the temperature extremes. The second model (non-linear temperature inhibition model) describes the inhibiting effect of either high or low temperatures on organism development. Sharpe and DeMichele propose a model which results in a linear response in mid-temperatures coupled with non-linear temperature inihibition at both high and low temperatures.

Sharpe and DeMichele developed a kinetic model that is based on the assumption that development is regulated by a single control enzyme whose reaction rate determines the developmental rate of the organism. Furthermore, they state that the control enzyme can exist in two temperature dependent inactivation states as well as in one active state. The first step in their model is the calculation of the probability that the control enzyme is in an active state. The only variable in this calculation is T, the absolute temperature. The development rate itself is described in terms of the control enzyme-substrate reaction. The final equation of Sharpe and DeMichele is rather complex and implies the knowledge of several constants reflecting the individual thermodynamic characteristics of the organism's control enzyme system which

16

is assumed to control development. In this model, temperature has two dominant effects upon enzymatic activity. The first is a result of a change in the distribution of molecular free energy with temperature, the so called rate reaction. The second relates to changes in enzyme activity. As the binding forces of the bonds which determine the enzyme structure and hence its activity are only somewhat greater than the thermal energies present in the environment, they are continuously being reversibly broken and reformed. Thus, the enzymatic activity resulting from these reversible transitions will show a critical temperature dependence (Sharpe and Hu, 1980).

This model was the first attempt to describe the reaction kinetics of temperature dependent development of poikilotherm organism including the non-linear parts of the development curve. As described above, the day-degree summation model has several disadvantages for the calculation of development rates, nevertheless, it is the model of choice when it comes to the calculation of larval age in forensic entomology as it is easy to use (see 1.5).

For calculating development rates it is also of importance whether experimental data was produced under constant or fluctuating temperatures. Studies of the effect of fluctuating temperatures on insect development often show that, given the same mean temperature, insects appear to develop at different rates in fluctuating conditions than they do at constant temperatures (Worner, 1992). Unfortunately, some studies show that development is accelerated under fluctuating temperatures, others claim it is decelerated. The differences between development predicted by nonlinear models under constant and variable temperatures for the same mean temperature was first described by Kaufmann (Kaufmann, 1932) and is called the Kaufmann effect or the rate summation effect: development will be faster at low temperatures and slower at high temperatures when comparing the development times measured under a constant temperature T_α to those measured at varying temperatures with a corresponding mean temperature equal to T_α (Tangioshi et al., 1976).

1.4.2 Nutrition

Growth and development of insect larvae is not only dependent on temperature. Numerous studies with poikilotherms indicate that an increase in nutrient quality increases the rate of development (Beck, 1950; Hobson, 1932; Kaneshrajah and Turner, 2004; Nijhout, 2008; Weissmann and Podmanická, 1970). The observed data indicate that the developmental rate is increased in proportion to the protein content of the diet. Sharpe and Hu (1980) analyze experimental data of the development of *Anthonomus grandis*. The larvae were reared on two different natural diets, one with a high amount of carbohydrates, the other with a high amount of proteins. They define growth as increase of biomass and development as the time between growth events or the duration of growth stages (Sharpe and Hu, 1980). Their results show that growth rate is highest on the high carbohydrate diet and developmental rate is fastest on the high protein diet. Sharpe and Hu propose an inverse relationship between rate of biomass increase and rate of development and state that protein content of the diet appears to be the primary driving force for development while carbohydrates appear to be the driving force for biomass increase.

Blow fly larvae mostly feed on a protein rich diet. However, in a study from Kaneshrajah and Turner (2004) it was concluded that the substrate the larvae were feeding on seems to matter when considering the rate of blow fly development. They determined the length of the larvae feeding on brain, heart, kidney, liver and lung at certain larval ages and showed that larvae feeding on lung tissue were nearly twice as long as larvae of the same age feeding on liver (Kaneshrajah and Turner, 2004).

1.4.3 Hormones

After hatching from the eggs, the developing blow fly maggots pass through 3 larval stages before they leave the nutrition site to enter the post-feeding stage and searching a proper space to hide and enter the last developmental stage: the pupal stage. In the pupal case the metamorphosis from larvae to adult fly takes place. During the feeding phases the larvae gain weight and size. As the cuticula can only stretch within limits and therefore allow growth only to a certain extent, for each new larval stage a moult is performed to replace the old cuticula with a new one. All changes during development are regulated by hormones. Shortly before the moult the ecdysone titer increases after a signaling of prothoracicotropic hormone (PTTH). PTTH is released by the corpora allata and directs the precise timing of the molt since its release is governed by both intrinsic factors such as size and extrinsic factors, such as photoperiod and temperature (Riddiford et al., 2003). The PTTH signaling leads to a release of ecdysone, the precursor of 20-hydroxyecdysone which induces the moult. As long as the juvenile hormone does not fall below a critical concentration the moult will lead to the next larval stage. If, however, the concentration of juvenile hormone is below a critical value, the next developmental step is pupariation and the initiation of metamorphosis. Once a larvae has reached a critical size and is committed to metamorphosis by a brief pulse of ecdysteroid release, it continues to feed. However, if such larvae were removed from the food source they would pupate and develop normally. Nevertheless, the resulting adult blow flies were significantly smaller in size but could still mate and reproduce. At the end of the feeding period the larvae leave the food and, if presented additional food, reject it. For some fly species the onset of wandering has a circadian basis. This is yet to be determined for most of the carcass associated blow flies. During the wandering phase larvae select the pupation site. As mentioned above, larvae characteristically display a strong photonegative response and usually burrow into the preferably dry substrate. A wet pupation site results in high mortality for many fly species. During the wandering period, food in the gut continues to be digested. Excess food is purged from the crop and replaced with a gas bubble (Fig. 1.6). Several environmental stimuli can influence the wandering period till a suitable place is found to start pupation. Shortly before pupation the larva shortens and resembles the shape of the later puparium but when disturbed it stretches again and seeks to avoid the interference. After final retraction of the anterior and posterior segments, the cuticle transforms into a hard puparial case and darkens gradually (Denlinger, 1994). Tanning is also controlled by 20-hydroxyecdysone as well as Bursicon. The onset of the pupation phase is controlled by a drop in the ecdysteroid titer. After it has dropped, the fly responds to the next ecdysteroid peak by initiating adult development. After the latter is completed, a well-defined circadian rhythm as well as the regulation by eclosion hormone determine the timing of eclosion.

1.4.4 Diapause

In addition to temperature the photoperiod is a crucial exogenous factor for the development of insects. The behavior of blow fly larvae can be altered when they are exposed to light conditions other than natural. Gomes et al. (2006) showed that larvae of *Chrysomya megacephala* burried twice as deep in preparation of pupation when exposed to 24 hours of light compared to the depth when exposed to a 12:12 light/dark cycle (Gomes et al., 2006).

Another important effect of the photoperiod on development is the induction of diapause. Diapause in blow flies was most studied for *C. vicina*. As it is a

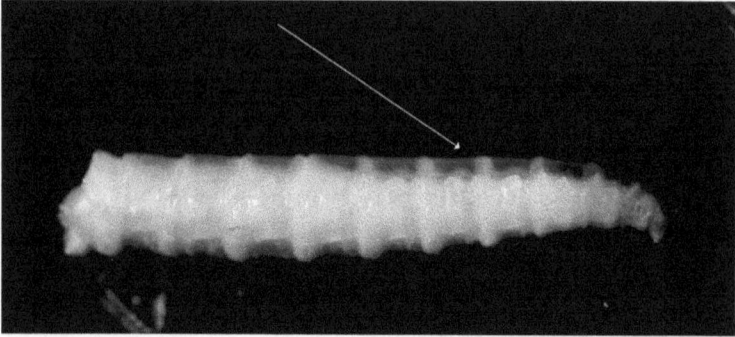

Figure 1.6: Post-feeding larvae: gas bubble replacing filled crop (arrow)

species adapted to cooler temperatures and therefore appearing on corpses in colder month like November or even December (Wetzel et al., 2009), it is very important to consider diapausing larvae in such month. The larvae of *C. vicina* enter a photoperiodically regulated diapause as a post-feeding larvae just before pupation (Saunders et al., 1986). The competence of the larvae to enter diapause is maternally controlled, the females themselves are influenced by the photoperiod they experienced before egg deposition (Saunders, 1987). Adult females exposed to long days lay eggs giving rise to larvae that develop to the next generation of flies without arrest. Females exposed to autumnal short days at temperatures below 25 °C lay eggs giving rise to larvae that enter diapause in the post-feeding stage (Saunders, 1987). The maternal critical daylength separating diapause-inducing short days from development-promoting long days is about 14.5 h of light per days (data for Scotland) (Saunders, 1987). Diapause commitment among the progeny of short-day females of *C. vicina* is also dependent on the temperature of the larval environment (Saunders, 1997). Diapause is only fully expressed when the maternally induced diapause competent larvae are exposed to temperatures below 15 °C (Saunders et al., 1986). If those larvae are exposed to higher temperatures during the post-feeding phase they will develop without arrest.

Case report

On the 4th of April 2008 a male corpse was found in his apartement. Doors and windows were closed, the heating was turned off. According to the police report

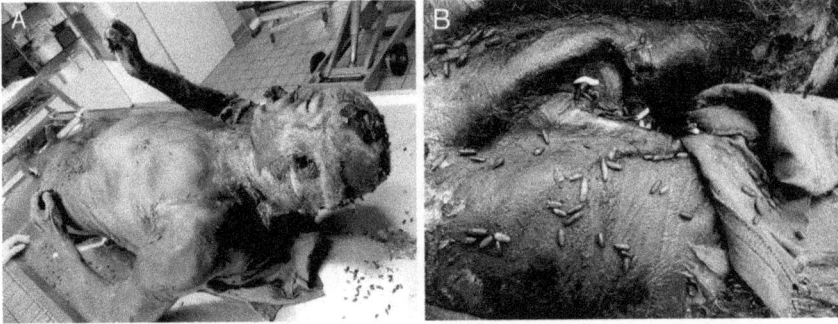

Figure 1.7: A: Mummified corpse, B: post-feeding larvae on mummified skin.

the temperature was about 18 °C. The body was lying in his bed in a mummified state of decomposition (Fig. 1.7 A) and hundreds of pupae of blow flies were on and under the bed sheets. The autopsy was performed 5 days after recovery of the corpse. During the autopsy adult blow flies hatching from the pupal cases could be observed. Furthermore, wandering, post-feeding larvae were found on the dried out corpse (Fig. 1.7 B). All specimens were determined as *C. vicina* (Rognes, 1991).

The person has last been seen in November 2007. It is assumed that adult blow flies deposited their eggs in late November or beginning of December 2007. As the photoperiod during that time was below 14.5 hours of light per day, it is likely that the eggs were maternally equipped with diapause inducing factors. As the heating was turned off, it is possible that the temperatures during december 2007 fell below 15 °C even inside the room. Therefore, it is very likely that the larvae feeding on the corpse diapaused in their post-feeding stage. When in March 2008 the days became longer and the temperature rose, the larvae finished their diapause, pupated and completed their development about the time of recovery of the corpse. That seems to be a possible explanation for discovering a mummified corpse which does not provide a nutrition site for larvae with post-feeding larvae and freshly hatching adult blow flies.

1.5 Estimation of post-mortem interval (PMI)

The term 'post-mortem interval' can be misleading when using it in combination with forensic entomology. With the help of the age of insect larvae feeding on a corpse it can be calculated how long the corpse has been infested with insects. However, a person can already be infested when he is still alive, e.g. a neglected person with severe wounds on a smoker's leg (Benecke et al., 2004) or as reported from soldiers in the first world-war, wounds that would have led to amputation when the blow fly larvae had not fed the necrotic tissue to clean the wounds (Büchner, 1965). In contrast, a person can long be dead and the larval age can still be a few days, when the corpse has been stored in a place with no access for the insects and has been put outside just a few days prior to the finding. Therefore, one must take into account every aspect before giving a statement about a post-mortem interval. In the latter case, if the insect evidence does not fit to the decomposition stage of the corpse, the police can use that piece of information and can additionally use the information about how long the corpse has been available for insects. However, from the insect evidence no information can be actually given about the PMI in that case. One has to be aware of the possible sources of error.

In 1.3.1 it was already pointed out that a PMI estimation can be done by relating certain species to decomposition stages in a timely manner and therefore getting a rough idea about the colonization time. Another method is to calculate the age of e.g. blow fly larvae feeding on the corpse and thereby giving a minimum time interval the person has been dead. This is only possible if the calculation is done for larvae of the first colonization wave. If the life cycle of the first larvae has already been completed and the adult flies started a new infestation of the corpse, it is only safe to calculate the time of the first completed life cycle. The basis of the calculation of the larval age is their temperature dependent development (see 4.2). At the same time this means that the temperature regime of the time and place of larval development has to be known. One method to approximate the temperature is to record the temperature at the desired place for a few days and compare them to data recorded at the nearest weather station. A regression analysis is applied to both data sets to get a formula that can then transform the data from the weather station covering the desired time frame into the temperature values that most likely influenced the developing larvae. The second requirement for the calculation of larval age beneath the temperature is the correct identification of the species (see 1.3.3) and the third is the determination of the developmental progress. That means determination of either the stage (see 1.4.3) or the length of the maggots is required, depending on the reference data set and the method that is to be used for the further calculation.

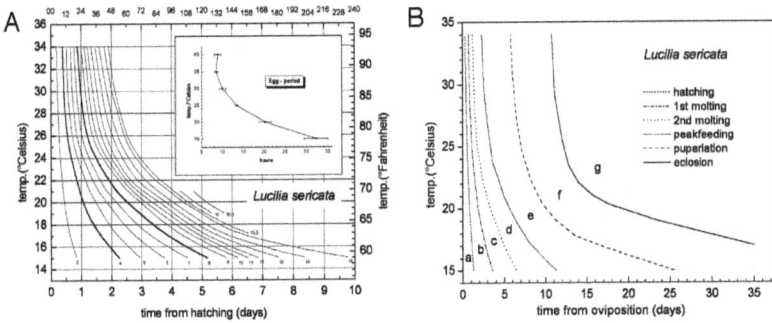

Figure 1.8: A: isomegalen Diagram and B: isomorphen Diagram for *L.sericata* from Grassberger and Reiter 2001.

In forensic case work, two different methods are frequently used to calculate a PMI. The first uses isomegalen or isomorphen diagrams, by which the lengths or the developmental stage of the larvae are combined as a function of time and mean ambient temperature in a single diagram (Grassberger and Reiter, 2001). According to its originators, this method is optimal only if the body and therefore the larvae were not undergoing fluctuating temperatures, e.g. in an enclosed environment where the temperature was nearly constant.

The second method of calculating a PMI estimates the Accumulated Degree Days or Hours (ADD or ADH). ADH values represent a certain number of "energy hours" that are necessary for the development of insect larvae. The degree-day or -hour concept assumes that the developmental rate is proportional to the temperature within a certain species-specific temperature range (overview in (Higley and Haskell, 2009)). However, the relationship of temperature and development rate (reciprocal of development time) is typically curvilinear at high and low temperatures and linear only in between.

The formula for calculating ADH is given by

$$\text{ADH} = T \cdot (\Theta - \Theta_0) \tag{1.2}$$

where T is the development time, Θ is the ambient temperature, and the minimum developmental threshold temperature Θ_0 is a species-specific value, the so called development zero, which is the x-intercept, i.e., an extrapolation of the linear

23

approximation of the reciprocal of time for development. This value has no biological meaning, it is the mathematical consequence of using a linear regression analysis (Higley and Haskell, 2009).

One basic condition for using the ADH method is that the ADH value for completing a developmental stage stays constant within certain temperature thresholds. For example a developmental duration for finishing a certain stage of 14 days at 25°C results in 238 ADD when a base temperature of 8°C is assumed. A developmental duration of 19 days at 21°C results in 231 ADD, both ADD-values are in the same range.

In general, the ADH method seems to give good results only when the larvae of interest have been exposed to temperatures similar to those used in generating the reference value applied in the PMI calculation (Anderson, 2001). Moreover, the temperature range in which the development rate is actually linear is not wide enough to cover all temperatures during a typical summer in Germany. Furthermore, neither developmental durations nor base temperatures for development have been calculated for species originating from Germany. The method must therefore be applied with caution.

Furthermore, it is highly problematic that uncertainties for temperature measurements from a crime scene cannot be taken into account in either model. It is difficult to determine the actual temperature controlling the larvae at a real crime scene. Since temperature is the variable that most influences development, it is crucial to consider it as accurately as possible. The standard procedure is to use temperatures of the nearest weather station for the desired time frame and correct them by applying a regression starting from temperatures measured at the crime scene, when taking the larvae as evidence (Archer, 2004a). The corrected values still contain uncertainties that cannot be accounted for by the methods currently used for PMI determination. No information exists for either model about the quality of the method or the error intervals of the calculated PMIs.

In general, the biggest problem is the lack of data for the development of certain species and especially data from different countries, as there is a geographical variation in thermal requirements for insect development (Honek, 1996).

1.6   Aims and questions of the thesis

The main purpose of the thesis was to apply the theoretical principles of forensic entomology to real forensic case work. Hence the choice of the Institute of Forensic Medicine, Bonn as working place. Three areas of the field of forensic entomology were picked to be improved by new experiments. Firstly, the ecological question of how the location influences insect colonization (chapter 2 and 3) was investigated on the basis of real cases. Experimentally the problem of how promptly blow flies oviposit on fresh carcass exposed indoors compared to outdoors was examined. A lot of corpses are found indoors and it was unreported if there is a delay in blow fly infestation. Therefore, an experiment was designed to compare in indoor and outdoor experiments the time interval between exposure of fresh piglet carcasses and the first deposited egg batches. Furthermore, the quantity of the egg batches was compared in both locations (chapter 4). Secondly, improved ways of insect identification based on sequence data were never tested for blow flies originating from Germany. Therefore, a published method using the COI-gene was tested for blow flies originating from Bonn (chapter 5). Thirdly, the problems in PMI estimation were tackled by designing a new model to calculate the PMI based on the actual growth behavior of blow fly larvae (chapter 6). The model is based on published data for the development of *L. sericata* from Austria and had to be validated with developmental data for *L. sericata* and *C. vicina* originating from Bonn (chapter 7).

1. Introduction

Part II

Impact of the location of a corpse for the estimation of a post-mortem interval

Chapter 2

Dumping of corpses in trash barrels - two forensic entomological case reports

2.1 Abstract

Two cases from the Ruhr Area in Western Germany are presented. In each case the deceased has been wrapped in plastic bags and placed inside a large garbage bin that was stored in the backyard of their properties. In both cases the relatives proclaimed a natural cause of death and the concealing of the deceased person ensured the further payment of the pension fund. In the first case those responsible stated a post-mortem interval of three years inside the bin, in the second case of half a year. Despite the closed lid of the bin the insect infestation was extensive and rich in species: empty pupal cases of several blow fly species were collected, as well as Histeridae and pupal cases of *Fannia scalaris* in the first case. In the second case Phoridae and larval skins of Dermestidae were additionally found.

2.2 Introduction

Two cases of a corpse placed in a organic waste collection bin by relatives are presented. In the first case the suspect stated a post-mortem interval of approximately three years, half a year was stated for the second. The evaluation of the entomological evidence was supposed to validate the statements of the relatives. Estimation of the post-mortem interval can be performed by calculating the age of the most developed larvae collected on the corpse (Catts and Goff, 1992). Shortly after death, blow flies (Calliphoridae) deposit their eggs on the corpse if the weather conditions are good and accessibility to the corpse is given. Pristine conditions for blow fly activity are an openly exposed corpse, no rain and temperatures about 25 °C (Norris, 1965). Is, however, the stimulus for oviposition exceedingly intense, the flies gain access even to carcasses in housings, wrapped and stored. The hatching first instar larvae develop temperature-dependent; between certain temperature thresholds development follows the rule the higher the temperature the faster the growth (Grassberger and Reiter, 2001). The larvae pass through three larval stages, feeding and growing constantly. After completing the third laval stage the larvae transfer to the post-feeding stage and leave the corpse to hide at a safe place and pupate (Reiter, 1984).Considering the developmental progress of collected larvae and the temperature they experienced during development at the place of discovery one can draw conclusions about the larvae's age and therefore about the minimal post-mortem interval (Goff et al., 1986). During a warm summer this method reveals reasonable results for about one month post mortem. A corpse exposed for a couple of months will be colonized successively by various species of flies and other arthropods in addition to the blow flies (Grassberger and Frank, 2004). The different species found on a corpse can indicate a post-mortem interval due to their distinct preferences concerning different stages of decomposition. Furthermore, some species can be an indicator for the season a corpse has been exposed.

2.3 Casuistry

2.3.1 Place of discovery

Case 1 (3 years post mortem)

In January 2005 the corpse of an eighty-year-old man wrapped in plastic bags was found inside a customary, 240 liters in volume, organic waste collection bin stationed in the backyard of a semi-detached house (Fig. 2.1A). The resident, a geriatric nurse, stated that she stored her father whom she had cared for until his death, inside the organic waste collection bin three years ago to continuously receive his pension and the money from the long-term care insurance. The corpse was discovered when the

2. Dumping of corpses in trash barrels - two forensic entomological case reports

Figure 2.1: Trash barrels used to disguise the bodies. A: Bin from Case 1 (the adhesive tape was attached while using the bin for a dead piglet). B: Bin from Case 2.

daughter could no longer avoid an overdue control visit regarding the care situation and handed herself in. The garbage bin was positioned in the backyard behind the attached garage and was surrounded by garbage and bulky waste. The general condition of house and backyard was messy.

Case 2 (1/2 year post mortem)

In November 2006 a forty-five-year-old man handed himself in at a police station in the Ruhr area and stated that he has stored the dead body of his eighty-five-year-old father inside the organic waste collection bin he was carrying along (Fig. 2.1B). He reported the bin including the body has been standing in the backyard since May 2006. Due to financial distress he had decided not to report the death but instead continuously received the pension of his father. The corpse was wrapped in plastic bags and stored inside the garbage bin standing in the backyard beside other bins. The backyard as well as the entire house were neat and clean.

2.3.2   Entomological findings

Case 1 (3 years post mortem)

31

2. Dumping of corpses in trash barrels - two forensic entomological case reports

Entomological evidence was collected from the corpse (Fig. 2.2 B) as well as out of the bin. Empty pupal cases of scuttle flies (Diptera: Phoridae) and blow flies (Diptera: Calliphoridae) were collected and identified. Furthermore, empty pupal cases of *Fannia scalaris* (latrine fly) were collected as well as adult specimens of hister beetles (Coleopera: Histridae).

Analysis:

As mentioned above, some blow fly species infest the corpse within a few hours post mortem. The empty pupal cases indicate that the developmental cycle from egg to eclosion of the adult blow flies has been completed. The required time frame for completing development is temperature dependent and can be completed within 3 weeks in the summer months (Grassberger and Reiter, 2001; Reiter, 1984). Nevertheless, Calliphoridae are not very active or present during winter time (Saunders and Hayward, 1998). As the corpse was supposedly put in the bin in December, blow flies won't have deposited egg packages until the next spring. Some species of Phoridae though are active during winter time and measure only a few millimeters in average, thus they are able to access a corpse even through tight openings (Disney, 2005). Both of these aspects, season and accessibility indicate an initial population by Phoridae since as soon as a corpse is dominated by larvae of Calliphoridae, there is almost no possibility for the much smaller Phoridae to also colonize the corpse.

The so called "latrine fly" *Fannia scalaris* is widely distributed through Central Europe, its occurrence is often associated with faeces and urin (Benecke and Lessig, 2001). The optimal development of the larvae proceeds in half-liquid faeces (Smith, 1986). The whole garbage bin smelled like faeces even after the body was removed. Apparently, *Fannia scalaris* also developed from egg to adult. Adults and larvae of hister beetles nourish predaciously on the blow fly larvae and pupae (Smith, 1986). They colonize a corpse right after the blow flies do, to devour their larvae. Every fly species found on the corpse produced at least one generation of descendants. These different successions indicate that the corpse had been stored in the bin for at least a couple of months, nevertheless regarding the entomological evidence it is impossible to state if the post-mortem interval has been a couple of years.

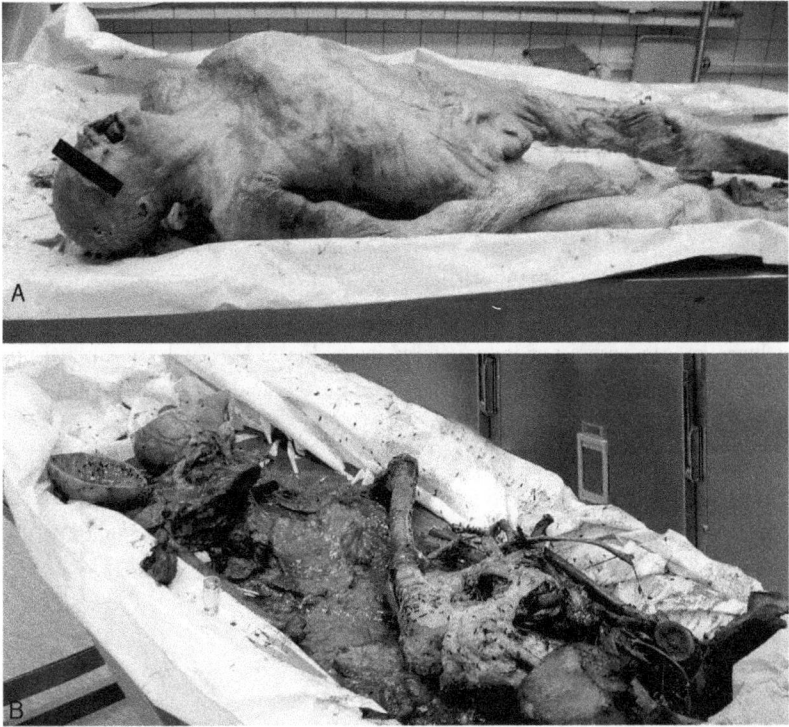

Figure 2.2: A: Body after recovery from bin in Case 2. B: Body from Case 1 after partialy performed autopsy

Case 2 (1/2 year post mortem)

During autopsy, entomological evidence was collected from the corpse (Fig. 2.2 A) and the garbage bin. Empty pupal cases of Calliphorinae, Luciliinae and *Protophormia terraenovae* (all blowflies) were found. In addition, adult specimens as well as pupal cases of Phoridae (scuttle flies), Fanniidae and more larval skins from *Dermestes lardarius* (skin beetles) were collected as well. Adults of *Necrophorus humator* (burying beetle) were fetched from the liquid on the floor of the bin.

Analysis:

The suspect stated that the corpse was put into the bin in May 2006. A confirmation for this statement from an entomological point of view is the finding of empty pupal cases of *Protophormia terraenovae* in the bin. Among the family of Calliphoridae this species is best adapted to cold (Grassberger and Frank, 2004). Peak temperature values for May 2006 were between 12 - 16 °C . The probability is high that the corpse was indeed stored in the bin in May and was infested primary by larvae of *Protophormia terraenovae*. Furthermore, the empty pupal cases indicate a completion of development.

Finding larval skins of *Dermestes lardarius* confirms that several month has passed since storage of the corpse in the bin. Megnin reports an appearance of the beetles 3-6 month post mortem, they feed on dry organic substances (Smith, 1986).

To sum up, the entomological findings in this case confirmed the statements of the suspect.

2.4   Discussion

Both presented cases concerning the disposal of corpses in a compost bin and the evaluation of collected entomological evidence show the advantages but also the difficulties in estimating a post-mortem interval by means of forensic entomology. During the analysis of the first case (3 years post mortem), the question arose how fast insects attain access to a corpse inside a bin. An experiment was conducted using the original garbage bin (Fig. 2.1 A) and exposing a dead piglet inside. The bin was kept shut for 2 weeks constantly. In contrast to the original case, the piglet was put in the bin in midsummer (July 2005) and not in December. Furthermore,

2. Dumping of corpses in trash barrels - two forensic entomological case reports

no packing material as plastic bags was used. The temperatures during the two weeks ranged between 20 and 25 °C. When opening the lid after two weeks, blow fly larvae in their 3rd instar stage were observed on the piglet as well as adult specimens of *Nicrophorus humator* in the liquid on the floor of the bin. The results of the experiment show that blow flies as well as large beetles will gain access to the corpse even in a closed bin probably by crawling through the venting slot which is simply a slot between lid and bin not tightly sealed. In contrast to the presented cases no adult or larval specimens of scuttle flies (Phoridae) were found. This is a major indication for the longer exposure times in both cases. In the experiment it was also shown that liquids from the corpse can not flow off from the bin. Thus, all liquid is collected at the bottom of the bin and can act as a trap for beetles and larvae. Furthermore, the liquid might promote formation of adipocere.

As both cases and the experiment show, insects will gain access to a corpse even in apparently difficult accessible environments as a closed bin and despite additional wrapping in plastic bags. Furthermore, despite the liquid on the bottom of the bin not all specimens drowned but managed to finish completing their life cycle, as shown by the emty pupal cases. Several specimens did not fall down when leaving the food source to find a suitable place for pupation but crawled in between the folded plastic layers and pupated there.

Interestingly, in both cases similar arthropod species were observed in the bins. A longer post-mortem interval inside the bin did not attract much more species despite the advanced stage of decay of the corpse. Therefore, it is not possible to estimate a longer post-mortem interval based on a species-rich succession of the corpse after several years.

To sum up, when a corpse is disposed inside a closed container as a compost bin with the tiniest opening, insects will gain access and colonize the corpse despite additional wrapping of the latter in plastic bags. Moreover, an estimation of the exposure time of the corpse inside the bin is possible within the first year but not after several years.

2. Dumping of corpses in trash barrels - two forensic entomological case reports

Chapter 3

Use of *Megaselia scalaris* (Diptera: Phoridae) for post-mortem interval estimation indoors

3.1 Abstract

In forensic entomology the determination of a minimum post-mortem interval often relies on the determination of the age of blow flies, since they are generally among the first colonizers of a corpse. In indoor cases the blow flies might be delayed in arriving at the corpse. If the windows are closed, the attracting odour is confined and does not reach the flies, so that it takes longer for them to find and access the corpse. If blow flies are delayed or are unable to reach a corpse lying inside a room, much smaller flies (Phoridae) can enter and deposit their offspring. Tthree indoor case scenarios are presented in which age determination of *Megaselia scalaris* gave much more accurate estimates of the minimum post-mortem interval than from larvae of Calliphoridae. In all cases the estimated age of the blow fly larvae was between 10 and 20 days too short compared to the actual PMI. Estimation of the PMI using developmental times of Phoridae can be a good alternative to the determination of blow fly larval age, since Phoridae are found inside apparently enclosed environments (sealed plastic bags or rooms with closed doors and windows) and also at temperatures at which blow flies are inactive.

Figure 3.1: Comparison of size of a blow fly (*L. sericata*)(shown above) and a phorid fly (*M. scalaris*) (shown below).

3.2   Introduction

Blow flies (Calliphoridae) are usually among the first visitors to a carcass. They use the carcass as nutrition for themselves or as a breeding site for their progeny (Norris, 1965). Determination of the age of the oldest blow fly larvae feeding on a corpse usually gives a good estimate of PMI_{min}. The earlier the blow flies arrive at the corpse soon after death the more accurate the estimate gets. Corpses are often found inside flats with closed windows, which causes a delay in blow fly infestation. In such circumstances a calculation of PMI_{min} might therefore lead to an underestimate of the actual PMI.

Scuttle flies (Phoridae) are much smaller than blow flies (Fig. 3.1) and can reach carcasses even inside closed plastic bags (Disney, 2008). Blow flies usually colonise a corpse rapidly, and their larvae produce large feeding masses that deter Phoridae from ovipositing in the same area. If Phoridae can reach the carcass first, however, they will oviposit immediately. Also, Phoridae are small enough to penetrate flats even when the windows are closed (Disney, 1994). Furthermore, they are active in winter when the temperature is too low for blow flies to be active (Manlove and Disney, 2008).

Three cases are presented in which age determination based on developmental data for *Megaselia scalaris* (LOEW, 1866) gave values much closer to the missing period of the deceased than values based on developmental data for blow fly larvae.

3. Use of *Megaselia scalaris* (Diptera: Phoridae) for post-mortem interval estimation indoors

3.2.1   Cases

All three cases were autopsied in the Instiute of Forensic Medicine in Bonn, Germany. The insect evidence was collected, preserved and identified. The insects were identified from morphological features, using identification keys (Disney, 1989; Rognes, 1991). Specimens belonging to the Phoridae family were also checked by another entomologist (Hans-Georg Rudzinski, Entomos), to ensure correct identification.

Case 1

Finding situation

A woman was found dead in her apartment in January 2009. She had not been seen for 40 days. She was lying on a mattress on the floor. The window in the room was closed and the shutters were down. All other windows and shutters in the rest of the flat were closed, apart from a tiny window in the bathroom. Next to her several pills were found. She was clothed in a T-shirt, thin capri-pants and woollen socks. Most of the body was mummified, and the face was partially skeletonized (Fig. 3.2 A and B).

Entomological evidence

Empty pupal cases of *Lucilia sericata* (MEIGEN, 1826) were found adhering to the woollen socks. A live specimen of *Necrobia rufipes* (DE GEER) was crawling out of the clothes during autopsy. Dead adult specimens were collected, as well as empty pupal cases of *Megaselia scalaris*. When the abdominal cavity was opened, thousands of larvae and pupae of *M. scalaris* were revealed (Fig. 3.2 C and D).

Conclusion

The temperature in the room when the corpse was found was 21°C . It was assumed that the temperature remained almost constant, as the windows were closed and the shutters were down. *L. sericata* takes 19 days to complete its life cycle (own developmental data; data of (Grassberger and Reiter, 2001) suggest 16 days). M. scalaris completes development at a 12:12 photoperiod after about 37 days (Trumble

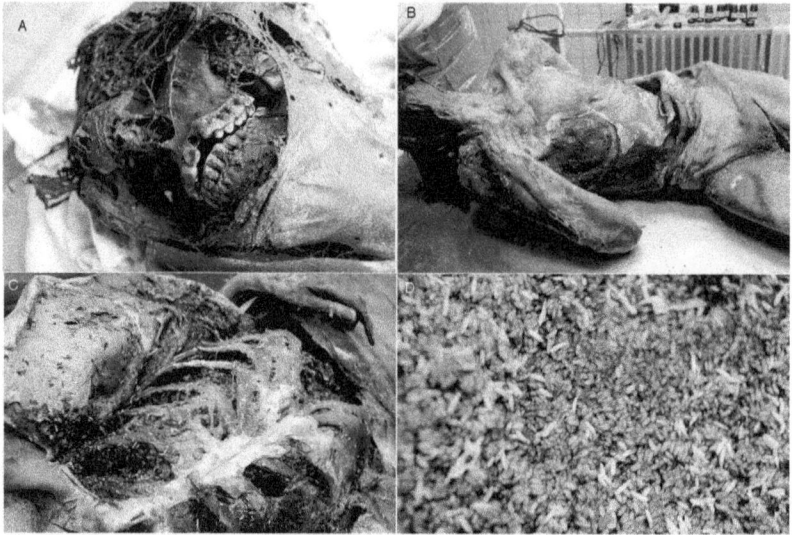

Figure 3.2: Case 1: A) and B) state of the corpse, C) open abdominal cavity, D) close up inside abdominal cavity showing hundreds of larvae and pupae of *M. scalaris*.

and Pienkowski, 1979). Since the individual's missing period was about 40 days, the PMI determination using developmental data of *M. scalaris* was much more accurate than the PMI calculated with data for *L. sericata*.

Case 2

Finding situation

The corpse of a man was found in his flat at the end of June 2008. He was lying in his bed, dressed in pyjamas. All windows in the flat were closed. The missing period was estimated as 36 days. The flat was described as neat. The corpse was in a state of advanced decay (Fig. 3.3 A).

Entomological evidence

Live specimens (freshly hatched) of *L. sericata* were collected during autopsy, as well as empty pupal cases. Furthermore, several pupae, empty pupal cases and dead

3. Use of *Megaselia scalaris* (Diptera: Phoridae) for post-mortem interval estimation indoors

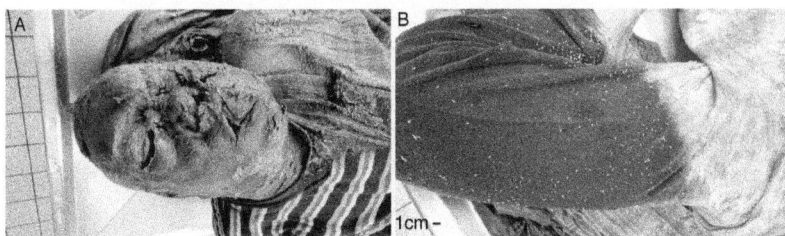

Figure 3.3: Case 2: A) state of decomposition, B) leg in pyjama with pupae of *M. scalaris*.

specimens of *M. scalaris* were found adhering to the fabric of the pyjama (Fig. 3.3 B).

Conclusion

A temperature of 20.5 °C was measured inside the flat when the corpse was recovered. Development of *L. sericata* takes about 19 days to complete at a uniform temperature of 21 °C (own developmental data; data of (Grassberger and Reiter, 2001) suggest 16 days). Under 16:8 light/dark conditions the development of M. scalaris at 21 °C is complete after 33 days on average (Trumble and Pienkowski, 1979). Again, the PMI estimation based on developmental data of *M. scalaris* fits the missing period better than an estimation based on data for *L. sericata*, assuming reasonably constant temperatures inside the flat.

Case 3

Finding situation

A male corpse was found in April 2008 in his flat 18 days after he has been seen for the last time. The windows were closed and the flat was described as messy. The corpse was in an advanced state of decay (Fig. 3.4 A).

Entomological evidence

Third instar larvae of *Calliphora vicina* (ROBINEAU-DESVOIDY, 1830) were

41

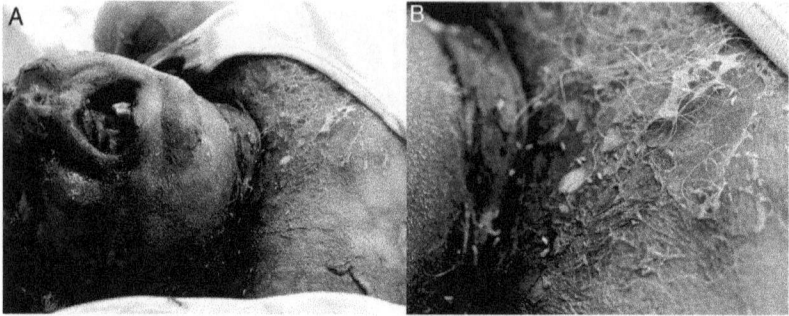

Figure 3.4: Case 3: A) Decomposition status of the corpse and B) pupae of *M. scalaris* in the chest area.

feeding on the corpse. They were collected and reared in a climate chamber at 25 °C. Hatched adults were first observed 11 days after the autopsy. A small number of larvae and pupae of *M. scalaris* were observed in the chest area of the corpse (Fig. 3.4 B). They were collected and preserved in alcohol.

Conclusion

Based on the temperature profile (21 °C inside the flat, 4 °C in the cooling device of the institute for forensic medicine, 25 °C in the climate chamber), the total developmental time for *C. vicina* was estimated as 23 days. After subtracting the developmental time in the institute (4 days cooling device, 11 days climate chamber), the development time in the flat was about 8 days (own developmental data for *C. vicina*). That is ten days less than the individual's missing period of 18 days. Larvae of *M. scalaris* pupate at 21 °C under 16:8 light:dark conditions after an average of 15 days, and post-puparition sets in after about 18 days (Trumble and Pienkowski, 1979). The pupae of *M. scalaris* therefore indicate a minimum PMI of 15 days.

3.3   Discussion

Several case reports consider the role of Phoridae in forensic cases (Campobasso et al., 2004; Disney and Manlove, 2005; Greenberg and Wells, 1998; Manlove and

Disney, 2008). Here a comparison of the post-mortem interval estimation based on developmental data for *M. scalaris* and blow fly larvae was presented. In all three cases the minimum post-mortem interval was far too short when it was based on developmental data for blow flies. This was due not to bad data or miscalculation but because it is difffficult for blow flies to infest corpses lying inside enclosed environments. Due to their size, blow flies take longer to a) recognize the smell and b) enter an enclosed room. Studies showed that even in cases where the window was ajar, blow flies arrived with a delay of 24 hours at small piglet carcasses exposed inside a room (Reibe and Madea, 2010a). Scuttle flies are much smaller, however, and can penetrate into or escape from apparently closed containers (Disney, 2008). In all three cases presented it can be assumed that the persons died shortly after they had last been seen. It has been shown that *M. scalaris* entered the flats shortly after death and began ovipositing. This led in all three cases to a reasonably accurate estimation of the PMI based on developmental data for *M. scalaris*. In general, all entomological evidence collected from the corpse shall be included in the PMI estimation and in indoor scenarios special attention should be directed towards phorid flies.

3. Use of *Megaselia scalaris* (Diptera: Phoridae) for post-mortem interval estimation indoors

Chapter 4

How promptly do blow flies colonise fresh carcasses? A study comparing indoor vs. outdoor locations

4.1 Abstract

It was investigated how long blow flies take to find and oviposit on fresh carcasses placed outdoors and indoors. Paired dead piglets, one in the open and the other in a nearby room (on the first floor of an occupied, detached, suburban house near Cologne, Germany, with a window opened 9 cm) were exposed simultaneously on nine occasions. The species visiting both locations and the number of egg batches deposited by blow flies between both locations were monitored 2, 8, 24 and 48 hours after exposure. In all cases the indoor piglet was exclusively infested by *Calliphora vicina*, only in one case, on a very hot day after 48 hours exposure did *Lucilia sericata* infest an indoor carcass. The outdoor piglets were infested by a variety of common corpse-visiting species: *Lucilia sericata, Lucilia caesar, Lucilia illustris, Calliphora vicina and Calliphora vomitoria*. A significant difference in the number of egg batches was detected between indoors and outdoors. Furthermore, in only two of nine runs did oviposition occur within the first 24 hours of exposure indoors. Ambient temperature, daylength and rainfall had no significant effect on the number of egg batches. Moreover, much less larvae were observed on indoor piglets; too few to form maggot masses. This might result in slower larval development than on outdoor piglets. It was concluded that PMI estimation for corpses found indoors must be handled carefully as oviposition might have taken place with a delay up to 24 hours.

4.2 Introduction

Corpses in houses or apartments are frequently found in late stages of decay infested by larvae of Calliphoridae, Phoridae, Muscidae or Sarcophagidae. Such persons are usually socially isolated, leading to delayed discovery of their bodies (Archer et al., 2005). In such cases a PMI determination is complicated since it is unclear how promptly the blow flies found the body and started laying eggs. Bodies can be colonized by insects in several different locations (Faucherre et al., 1999) including poorly accessible environments (Goff, 1991a). To estimate the colonization period, it is therefore important to know how soon the insects can obtain access. Insect-infested bodies with a known post-mortem interval, investigated in the institute of forensic medicine, Bonn serve as control cases for PMI determination using insect larvae. Unfortunately, for bodies found indoors the presumed post-mortem interval often ranges between days and weeks as nobody missed the person immediately after death. Thus, it is difficult to verify an estimated post-mortem interval for corpses found indoors. In most cases examined, the time when a person was last seen and the estimated larval age differed by several days, possibly because the person did not die until several days after they were seen or because the blow flies needed considerable time to perceive the body and enter the apartment. An experiment was designed to observe the time interval between exposure of a body inside a room of a house with a slightly open window and first colonization by blow fly eggs.

4.3 Material and Methods

Dead piglets exposed in the open were compared to piglets lying inside a 20 m² room with a window opened 9 cm at the upper edge, allowing a total entry area of 0.18 m² (Fig. 4.1A). The room was on the first floor of an occupied, detached house with a garden, located in a suburb of Cologne, Germany, and was fully equipped with a bed, bookshelves and a closet. No food or organic waste was kept in the room. The

4. How promptly do blow flies colonise fresh carcasses? A study comparing indoor vs. outdoor locations

Figure 4.1: A. Room on first floor in detached house as indoor location with tilted window (arrow) facing north-west. B. Outdoor location with table and basin containing the piglet in front of a garden shed with a small eave. Picture taken at 1:00 p.m, showing the sun and shading situation.

piglets were placed in a plastic basin on the floor in the middle of the room. The window faced north-west, so that the sun shone through it between about 3:00 and 6:00 p.m. The light was switched off during the experiment and nobody entered the room except to check the piglets at distinct time intervals. The heating was turned off. The control piglets were placed outdoors about 50 meters from the window, in the garden on the south side of the same house, directly in front of a garden shack with small eaves (Fig. 4.1B), to ensure that they were sheltered from extreme rainfall. Additionally, they were exposed to the sun in the same manner as the indoor piglets: direct sunlight reaching the piglets at late afternoon. These piglets were also placed in a basin and on a table to prevent vertebrates from disturbing them.

In total, nine experimental runs were carried out successively, using altogether 18 dead piglets; nine indoor and nine outdoor. Run 1 was started at the end of May 2008, Runs 2, 3 and 4 were evenly distributed during June 2008, Runs 5, 6 and 7 in July 2008 and Run 8 and 9 successively in August 2008.

The piglets were always exposed in pairs, one indoors and the other outdoors simultaneously. They each weighed between 1 and 2 kg and were frozen at -20°C and defrosted within 24 hours before exposure. Shortly before exposure they were cleaned with running water. Each experiment started between 11:00 a.m. and 1:00 p.m., and stopped approximately 50 hours later. The ambient temperature was recorded by *Ebro EBI-6* Dataloggers lying directly next to the piglets, rainfall was noted and the amount of rain was taken from a private weather station

47

4. How promptly do blow flies colonise fresh carcasses? A study comparing indoor vs. outdoor locations

(http://www.koelschwetter.de) in a nearby village (about 3.7 km away).

The piglets were checked 2, 8, 24 and 48 hours after exporsure, taking high resolution pictures using a *Panasonic Lumix DMC-TZ2* digital camera. Pictures of every body part and the natural orifices of the piglet were taken to record the presence and the number of egg batches deposited by blow flies. An egg batch was defined as a cluster of eggs probably oviposited by one female. The egg batches were counted on a computer screen using *Adobe Photoshop* to enlarge all pictures in the same manner. When discrete egg batches could not be distinguished due to several females contributing to one egg batch, a digital circle the size of one egg batch (taken from a position where a batch was distinguishable) was used as an on-screen template to determine the approximate number of egg batches. Additionally, although after 48 hours of exposure larvae had hatched, - and so technically could not be counted as egg batches, for statistical purposes the number of new egg batches that were deposited between 24 and 48 hours of exposure was added to the number counted after 24 hours.

To determine the species visiting the piglets, adult flies (if present) were caught either by hand (by carefully lowering a small glass jar on the fly) or using sticky traps (*Aeroxon*, 21x6 cm, without attractants or toxic substances) for approximatley 30 minutes. Indoors, adult flies were only caught if eggs had been deposited already. Feeding larvae were not identified as it was not the primary question of the experiment which species actually infested the carcasses.

Statistical analysis was performed using *SPSS*. The count of egg batches after each of the four time intervals on piglets exposed indoors vs. outdoors was analyzed using a Mann-Whitney U-test and effect sizes (a measure of the strength of the relationship between two variables) for the impact of the location were calculated. A chi-square test was performed to test for differences in the composition of the captured species indoors and outdoors. Additionally, the location, the exposure time, the mean temperature, the daylength and the amount of rain were correlated to the number of egg batches to investigate the influence of these parameters.

4.4   Results

During the nine experimental runs the daily mean temperature ranged between 17.8 °C and 23.9 °C. The mean temperature value for all runs was 19.9 °C outdoors and

4. How promptly do blow flies colonise fresh carcasses? A study comparing indoor vs. outdoor locations

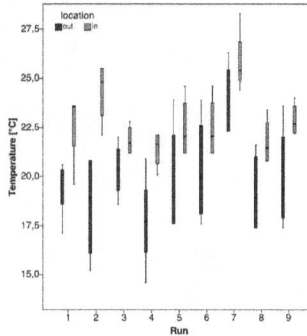

Figure 4.2: Temperature regime for each of the experimental runs conducted between May and August 2008.

Table 4.1: Mean number of egg batches indoors vs. outdoors at distinct exposure times

hour	location	mean count	SD
2	out	18.56	14.49
	in	0.22	0.67
8	out	55.56	39.41
	in	0.67	1,41
24	out	104.44	50.53
	in	2.5	1.19
48	out	132.22	48.42
	in	13.5	17.99

22.5 °C indoors. The temperature profile for all runs are shown in Fig. 4.2, the corresponding total rainfall in Fig. 4.3 A and the individual rainfall for each run in Fig. 4.3 B.

4.4.1 Number of egg batches

The number of egg batches was counted after 2, 8, 24 and 48 hours of exposure time (Table 4.1). The number of egg batches on the outdoor piglets was significantly higher for all exposure times than on the indoor piglets and the latter were infested

49

4. How promptly do blow flies colonise fresh carcasses? A study comparing indoor vs. outdoor locations

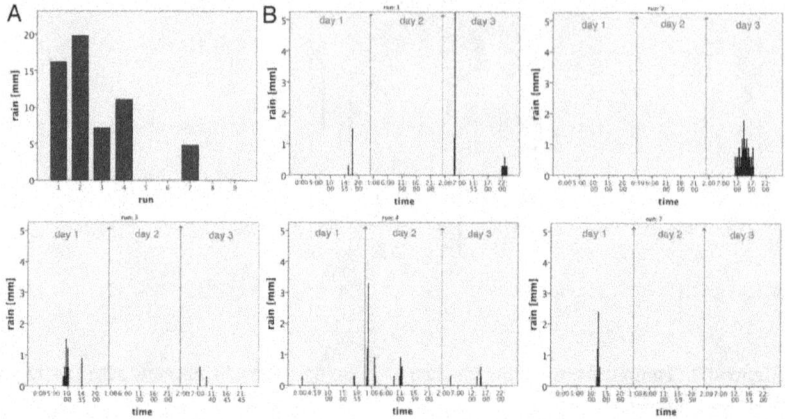

Figure 4.3: A. Sum of rainfall for each run. B. Individual rainfall profile for each run where rainfall was recorded, arrows indicate midnight.

later. Only in two of the nine runs did egg batches appear within the first 8 hours of exposure on indoor piglets. The mean number of egg batches was distributed as shown in Table 4.1. The standard error for 48 hours indoors is large because - although the median for the count was 6 batches there was an outlier in Run 7, which had the highest indoor temperature: 27 °C (Fig. 4.2). On that particular day, after 48 hours of exposure, there were about 40 egg batches on the piglet and about 20 adult flies in the room, while in each of the other runs there were only 5-10 egg batches.

The number of egg batches was significantly different between locations (using Mann-Whitney U-test $p \leq 0.001$ in all cases). The effect sizes for each exposure time ranged between r=0.84 and r=0.87.

Fig. 4.4 shows exemplarily the number of egg batches and rate of decomposition for the two piglets in Run 8. It is obvious that the number of egg batches on the outdoor piglet is already higher after 8 hours exposure time than on the indoor piglet after 48 hours. Moreover, due to the higher number of hatched maggots on the outdoor piglet, an advanced state of destruction was observed after 48 hours compared to the indoor piglet.

50

4. How promptly do blow flies colonise fresh carcasses? A study comparing indoor vs. outdoor locations

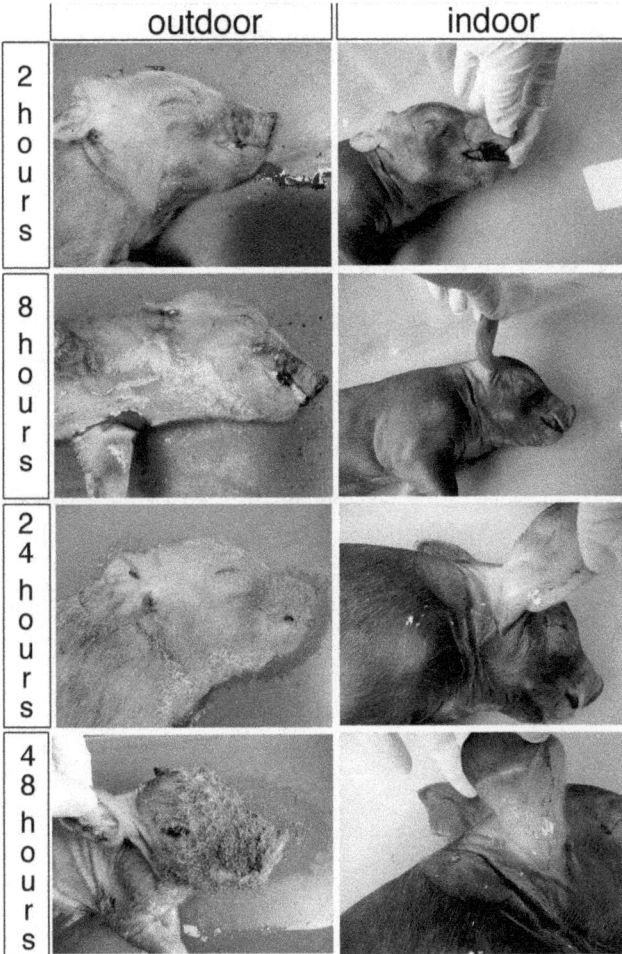

Figure 4.4: Amount of egg batches and decomposition state for piglets exposed outdoors (left column) and indoors (right column) 2, 8, 24, and 48 hours after exposure.

Table 4.2: Percentage distribution of the species caught indoors vs. outdoors during
all runs

Species	indoors	outdoors
C. vicina	75%	14.8%
C. vomitoria		1.6%
L. caesar		63.9%
L. illustris		4.9%
L. sericata	25%	14.8%
total number	12	61

4.4.2   Species identification

The majority of the Calliphoridae species caught on the outdoor piglets were identified as *Lucilia caesar* (LINNAEUS, 1758) (Table 4.2). The second most recorded
species were *Lucilia sericata* (MEIGEN, 1826) and *Calliphora vicina* (ROBINEAU-
DESVOIDY, 1830). The least recorded species recorded were *Lucilia illustris* (MEIGEN,
1826) and *Calliphora vomitoria* (LINNAEUS, 1758). The species caught indoors was
almost always *C. vicina* (in low numbers) except in Run 7 where after 50 hours of
exposure, up to 20 individuals of *Lucilia sp.* were observed but only 3 could be
caught by hand and identified as *L. sericata*. The numbers in Table 4.2 represent
counts of the specimens caught only (indoors: n=12, outdoors n=61); for the indoor
piglet they represent the number for *C. vicina* present quite accurately. In contrast,
the outdoor piglet was visited by many more individuals that evaded capture. Other
insect families appeared on the piglets too but only Calliphoridae were surveyed, as
they are the most common group of early corpse infestation insects. The chi-squared
test showed that the composition of the caught species indoors and outdoors was
significantly different (p ≤ 0.001).

4.4.3   Correlations

The number of egg batches was highly correlated to the location (rs=-0.82, p≤0.001)
and to the duration of exposure (indoors rs=0.81, p≤0,001, outdoors rs=0.80,
p≤0.001), but not to temperature (rs=0.16, p=0.35), rainfall (rs= 0.03, p=0.85) or
daylength (rs=-0.37, p=0.83) neither indoors nor outdoors. The only temperature-
dependent effect ocurred during Run 7, when the indoor temperature increased to
27 °C and many adult *L. sericata* flew in and deposited a large number of egg

batches.

4.5 Discussion

4.5.1 Delay in oviposition

It has been reported that blow flies can detect carcasses inside buildings and enter
dwellings to oviposit (reviewed in (Anderson, 2001)). Furthermore, bodies found
wrapped or inside a vehicle or trash can have been reported to be infested by fly
larvae (Goff, 1991a; Reibe et al., 2008). For these cases it is not yet established how
much time has to be added to the estimated larval age to approach a realistic PMI.

The present study was designed to detect the general magnitude of any differences
in the colonization rate of carcasses by blow flies relative to the location of the
carcass: outdoors or indoors. The number of egg batches in both locations differed
considerably throughout the first 48 hours, being significantly less indoors in all
runs. Although the method of counting the egg batches might be suboptimal when
several females contributed to a cluster of eggs or at other occasions just spread
a few eggs, the aim of the study was fulfilled satisfactorily as the results does not
depend on 10 egg batches more or less. Location therefore had a large effect on
the number of egg batches. Additionally, the indoor piglets were colonized up to
24 hours later than the outdoor piglets. In only two of nine runs (Run 3 and 4)
was the indoor piglet infested the same day as it was exposed. Although carcass
size does not matter in relation to arthropod succession patterns (Hewadikaram and
Goff, 1991) there might be an effect inside rooms of odours building up faster when
the carcass is larger and therefore attracting more flies faster. This is yet to be
inverstigated.

In addition, the different number of egg batches between carcass exposed indoors and outdoors resulted in a different amount of larvae. Greater larval activity
leads to increased tissue destruction and the formation of maggot masses in which
temperatures up to 20 °C higher than the ambient air temperature can be reached
(Slone and Gruner, 2007). The few larvae on the indoor piglets did not produce such
aggregations, so that larvae may develop more slowly indoors than outdoors, even
though indoor temperatures were about 2°C warmer than outdoors. This must be
taken into account as a source of uncertainty in PMI estimates when corpses found

indoors are infested with only small amounts of maggots.

4.5.2 Effects of temperature

It was shown that the temperature profiles of each of the experimental runs during summer time had no significant effect on the arrival of blow flies at the exposed piglets (indoors and outdoors) or the number of egg batches. However, this certainly only applies in months in which the temperatures are equal to the experimental period. Two additional runs in winter were included, one in November (mean temperature: 6 °C) and one in December (mean temperature: -2,7 °C) 2008. In both runs no colonization by insects on the indoor piglets was observed. In November the temperature raised to 14.9 °C on day 9 of exposure and blow flies oviposited on the outdoor piglet. In December no fly activity was recorded. Hence, temperature does matter when it comes to the general question of insect activity due to low temperatures. The results therefore apply only at moderate temperatures in the average range of about 15 °C and 25 °C. The absence of adult flies on the outdoor piglet was recorded whenever the piglet was temporarily exposed to extreme sunlight and very high temperatures (about 30 °C). The results do not reflect this observation as extreme weather situations are compensated by the choice of distinct exposure times. Even if there was one hot and sunny hour between 2 and 8 hours of exposure, it had no effect on the total number of egg batches after 8 hours.

4.5.3 Effects of rain

Rain did not prevent the flies from ovipositing. Either they waited for a dry moment or they could reach the carcass by leaving the place of shelter and crawl towards the carcass. As seen in the individual rain profiles (Fig. 4.3) in each run dry periods have been recorded. So far, little is known about the effect of rainfall on the oviposition behaviour of blow flies. Archer (Archer, 2004b) discovered that higher temperatures and rainfall increased both mass loss rate and decomposition stage progression rates of exposed neonatal remains. Nevertheless, no mention was made of whether flies approached breeding sites or laid during rainfall. Furthermore, it has yet not been reported if egg batches might be washed away by heavy rainfall. The observations show no correlation between rainfall and oviposition, hence even if it is raining the flies will oviposit the same day as exposure of the outdoor carcass. The eaves of

the roof sheltered the outdoor piglets from heavy rainfall and although rain reached
the piglets, eggs were not drowned completely. It was observed that whenever rain
reached the carcass, the feeding maggots crawled beneath the piglet and fed from
the moist tissue. No washing away of egg batches or maggots was recorded.

4.5.4   Blow fly species entering houses

Although no general correlation between air temperature and the number of egg
batches was observed, either indoors or outdoors, once, when the temperature indoors rose up to 27 °C after 45 hours of exposure, the piglet bloated heavily, many
adults of *L. sericata* arrived and about 40 egg batches appeared. It seems the high
temperature, extreme sunlight and enhanced smell associated with bloating stimu-
lated the flies to enter the room and oviposit. It has been reported that *L. sericata*
is usually found in bright sunlight (Schumann, 1971), while *L. caesar* is associated
with shaded places. The dominant species at the outdoor piglet was *L. caesar* as the
piglet remained in shade for three quarters of the day (see Materials and Methods).

Except in Run 7, *C. vicina* was exclusively caught at indoor piglets. This species is
known for favoring shady situations and urban habitats (Erzinclioglu, 1996; Horenstein et al., 2007). The entering of *C. vicina* might therefore be related to the
orientation of the window of the one house used in the experiment. It has to be
investigated if a window orientated to the south has different effects on the arriving
species. In 34 cases handled by the institute of forensic medicine, Bonn in 2008 *C.
vicina* was associated with 27.3 % of the indoor cases; in 21.2 % no insects were
found; in 24.2 % other *Calliphorinae* (especially *Luciliini*) were collected; in 15.2 %
the species remained unidentified; and in 12.1 % other families of flies like Phoridae and even beetles were observed (unpublished data). In these cases there was
a tendency to find *luciliine* species whenever the place of discovery was described
as messy and neglected. It would be interesting to observe the time till oviposition
when fresh carcasses are exposed in rooms containing additional organic waste.

A study from Germany showed that *L. sericata* had the highest synanthropy
index of all captured blow fly species (Steinborn, 1981), another German study
showed that *L. sericata* and *C. vicina* were the only blow fly species caught indoors
(Schumann, 1990). Moreover, in the sonducted study an exposure period longer
than 48 hours could have led to accumulation of *Luciliinae*. Nevertheless, these
findings indicate that in a clean room, facing north-west the first species to find a

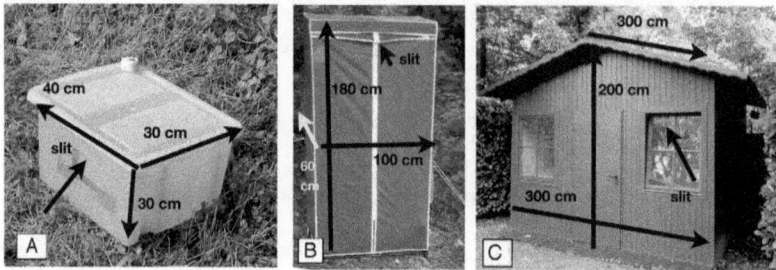

Figure 4.5: A) Small box used as a model for indoors, arrow shows the window-like
opening (10 cm long slit). B) Large box 1 (locker): 54 times larger than the volume
of the piglet, the slit is indicated by an arrow. C) Large box 2 (garden shack), 900
times larger than the volume of the pig, the tilted window is indicated by an arrow.

carcass and oviposit is *C. vicina*.

4.5.5 Conclusion

In summary, indoor carcass may be infested with blow fly egg batches up to 24 hours
after exposure. Generally, very few blow fly individuals enter the room to oviposit,
resulting in a significantly lower number of egg batches than on outdoor carcasses.
This might result in slower larval development as the smaller larval aggregations
would generate less metabolic heat. All these factors have to be taken into account
when a PMI(min) has to be determined for corpses found indoors.

4.6 Preliminary experiments for finding a suitable indoor model

Before the indoor experiments were conducted in a real house, other possibilities for
an indoor model were tested. The results are shown below.

4.6.1 Experimental design

In total 18 dead piglets in different location types were compared. Piglets exposed
openly (n = 10) were compared to piglets that were kept indoors (n = 8). Indoors
refers to containers of different types with distinct sizes and materials (Fig. 4.5)

4. How promptly do blow flies colonise fresh carcasses? A study comparing indoor vs. outdoor locations

that were used as models for a flat or normal house. The indoor models were differentiated in small (n = 4) (Fig. 4.5 A) and large (n = 4) (Fig. 4.5 B, C). The volume of the small boxes was about 20 times larger than the volume of the piglets itself, the volume of the large boxes was 54 or 900 times larger than the volume of the piglets. The small boxes (smbox) had a size of 40 cm x 30 cm x 30 cm and a volume of 40 L, with an opening simulating a tilted window (10 cm length, see Fig. 4.5 A). Two different types of large boxes were used, a locker (180 cm x 100 cm x 60 cm; 1080 L) (Fig. 4.5 B) made of fabric and a garden shack (300 cm x 200 cm x 300 cm; 18000 L) made of wood (Fig. 4.5 C). The piglets put in the small boxes had a volume of 2 L, the ones put in the large boxes a volume of 20 L. The pigs were deep-frozen the same day they died and were defrosted in a plastic bag 24 hours before exposure without flies having access to it. The piglets were exposed several times in the summer 2006 and 2007, sometimes only one pig, open or covered, sometimes both situations close together as well as 50 m apart from each other. The results from the experiments with openly exposed piglets (n=10) were pooled as well as the results for the piglets in small boxes (n=4) and for the piglets stored in large boxes (n=4).

The temperature data was recorded using a data logger (EBI-6, Ebro) and als the weather (e.g. sunny, cloudy, rainy) for the exposure days was noted on a data sheet. The piglets were checked daily and the egg batches were counted after taking macro high resolution pictures every day at the same time (late afternoon). Statistical analysis was performed with the program SPSS using the Kolmogorov-Smirnov Test for significance. A value of $p < 0,05$ was considered to be significant. Kolmogorov-Smirnov Test acts similar as the Mann-Whitney U Test, but is used for small sample sizes (Field, 2005). The Spearman Correlation factor rs shows the correlation between location type (open, smbox and lbox) and the 3 investigated variables (number of eggbatches, first day of egg deposition, number of egg batches after 24 hours) where -1 is a perfect negative correlation and +1 a perfect positive correlation.

4.6.2 Results

Weather and ambient temperature

The temperature for the exposure days ranged from 20°C - 25°C , with two exceptions. One day 31°C were measured whilst exposure of the piglet and one time only 18°C . The weather was always sunny except for one day, where it rained, when a piglet was exposed openly.

On which day have the first egg batches been deposited by blow flies?

For each location it was observed on which day after exposure of the piglets the first egg batches were present. The exposure day of the piglets equals day 1 (zero hour) of the experiment. When the piglets were exposed openly, the first eggs were observed during the first 24 hours in every case (Fig. 4.6: 1.a) and 2.a)). The shortest time interval observed for oviposition was 50 minutes after piglet exposure. Even during a day full of rain, egg batches could be observed within 24 hours but not as many as on a sunny day (data not shown). The results for piglets lying open compared to such lying in small boxes were very similar (Fig. 4.6: 1), Fig. 4.6: 2)). In both cases (open and small box) the day of first observation of egg batches is day 1 or 2 (that is within 24 hours), so there is no significant difference between both locations (p=0,99) (Fig. 4.6: 1.a)). Inside large boxes egg batches were never observed on the same day as piglet exposure. Also noted was a large variance in the time period until first eggs are deposited (Fig. 4.6: 1.a)). In one experimental run no eggs were recorded until day 7 after exposure, although several adult flies were present around the large box and even landed on it. Nevertheless the results showed no significant difference between open exposure site and large box concerning the day the first eggs were deposited: p=0,08.

How many egg batches were counted 24 hours after exposure of the piglets?

On openly exposed piglets the number of egg batches 24 hours after exposure was on average 61 with a standard error of the mean of 9.4. No significant difference to the results of the small box experiments (p=0,35) was observed. As to be seen in Fig. 4.6: 1.b) there is only a tendency toward fewer egg batches on piglets put in small boxes. In both cases there was a minimum of ten egg batches 24 hours

4. How promptly do blow flies colonise fresh carcasses? A study comparing indoor vs. outdoor locations

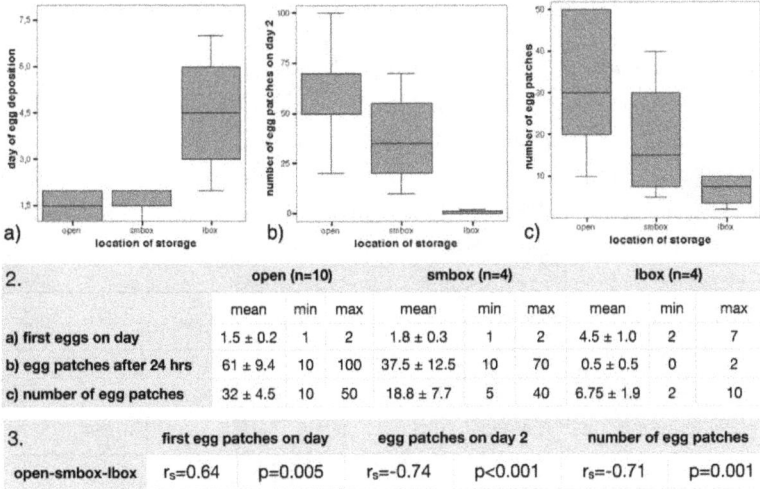

2.	open (n=10)			smbox (n=4)			lbox (n=4)		
	mean	min	max	mean	min	max	mean	min	max
a) first eggs on day	1.5 ± 0.2	1	2	1.8 ± 0.3	1	2	4.5 ± 1.0	2	7
b) egg patches after 24 hrs	61 ± 9.4	10	100	37.5 ± 12.5	10	70	0.5 ± 0.5	0	2
c) number of egg patches	32 ± 4.5	10	50	18.8 ± 7.7	5	40	6.75 ± 1.9	2	10

3.	first egg patches on day		egg patches on day 2		number of egg patches	
open-smbox-lbox	r_s=0.64	p=0.005	r_s=-0.74	p<0.001	r_s=-0.71	p=0.001

Figure 4.6: Locations: open= openly exposed piglets, smbox = piglets put in small boxes, lbox = piglets put in large boxes, a) day of first ovipostion, b) number of egg batches 24 hours after exposure (day 2), c) number of egg batches on the first day eggs were deposited. 1. Boxplots. 2. Descriptive statistics. 3. Spearman's correlation coefficient rs and associated p-values for the order of the locations open to small box to large box .

after exposure (Fig. 4.6: 2.b)). In contrast the use of large boxes resulted in severe differences between outdoor and indoor locations concerning the number of egg batches (p=0,007). In three of four cases no egg batches on the piglets were observed, in the fourth case two egg batches were counted after 24 hours exposure time inside a large box.

How many egg batches were counted during the first oviposition day?

On openly exposed piglets an average of 32 egg batches was counted, in the small boxes about 18 egg batches and in the large boxes an average of 7 egg batches on the first day eggs were observed (Fig. 4.6: 2.c)). As to be seen in the boxplot (Fig. 4.6: 1.c)) the trend from open to smbox to lbox for the amount of egg batches is decreasing. The p-values for the differences between the locations are: open vs. small box p=0.61, and open vs. large box p=0.02. Again, a significant difference can only be sen between openly exposed piglets and such put in large boxes.

4. How promptly do blow flies colonise fresh carcasses? A study comparing indoor vs. outdoor locations

Correlation

The impact of the location on the three variables mentioned above was tested with Spearman's correlation coeffizient rs (Fig. 4.6: 3). For the elapsed time until the first eggs were deposited the correlation factor is 0.64 with a significance value of p = 0.005, showing a strong positive relation between the location and the day the first eggs are deposited. Positive relation in this case means: later. The larger the box, the later the flies deposit their eggs. For the numbers of egg batches 24 hours after exposure of the piglets Spearman's correlation coeffizient is rs=-0.74, showing a negative relation between the location and the number of egg batches, which is strongly significant (p < 0.001). Negative relation here means: fewer. Fewer eggs were counted the larger the indoor location got. The same applies for the number of egg batches on the first day of oviposition: rs=-0.71. The larger the box, the less egg batches were present inside at that time (p = 0.001).

4.6.3   Discussion

The purpose of the study was a transfer of the experimental situation to a scenario when a human corpse is found indoors and additionally the application of the preliminary results to real case work.

The results of the experiments show definitely a trend when it comes to evaluate post-mortem intervals for corpses found in an enclosed environment. There is a relation between the location of storage of a corpse and the resulting colonization by blow flies, shown by the Spearmann's correlation. A dead piglet in an empty and clean large box or room was not colonized by blow flies the same day it has been exposed - in contrast to an openly exposed piglet, that was found by blow flies within 24 hours post exposure in all observed cases.

Only for the investigated variable of the elapsed time until the first oviposition no significant differences can be seen between the results for the open site and the large box exposure site. Though the value was very close to the real significance limit, no safe time period can be postulated in which the flies find their way indoors, only a trend can be shown towards a declaration of the colonization time. Nevertheless, the results from the Spearmann correlation give a significant difference as well as a a strong correlation for the time until the first egg batches are deposited and the size of the location. This may be due to the chosen statistical test and the small sample size of these preliminary results. The fact that the small boxes as a placement of dead piglets did not restrain the flies from depositing their eggs within 24 hours can be a helpful result for the investigation of dead infants hidden in bags or boxes.

One reason for the differences between small and large boxes might be due to an elongated time the odour might stay inside the large room and not reaching the flies outdoors. The smell from piglets inside a small box was recognizable even for the observer, when he was standing next to it. For the future repetitions of the experiments with only one kind of indoor scenario (a real house) are planned as well as an enlargement of the sample size to get more data. However it is not possible to wipe out the problem that the weather will never be exactly the same as the trial before and the location is also one of a kind and might never be similar to another scene concerning all its microclimatic influences. Another point of interest during further experiments will be the variation in insect species colonizing the bodies. A study from Hawaii showed that some species were restricted to remains discovered indoors, and also a greater variety of Diptera larvae were associated with them, while remains discovered outdoors had a greater variety of Coleoptera species present (Goff, 1991b). Coleoptera species were only recoreded at outdoor locations (data not shown) so in further experiments the species that colonize a body indoors and outdoors will also be determined. It can be concluded that for bodies found indoors, the PMI determination should be conducted with care as it is yet to be determined how the colonization time varies compared to bodies lying in the field.

4.7 Final remarks

As to be seen in the preliminary experiments that lead to the experiments conducted in a real house, the trends of the results were indeed visible. Nevertheless, no model was good enough to simulate a real indoor scenario. However, an important result of the preliminary experiments was the colonization behavior of blow flies towards piglets exposed in rather small containers. As it was shown, these enclosing did not prevent the blow flies from ovipositing shortly after exposure. As already mentioned in the discussion, this can be a helpful result when infants are found outdoors but inside a box or the like. Furthermore, it was a good experience to develop the experimental design further after ruling out the flaws of the preliminary experiments.

4. How promptly do blow flies colonise fresh carcasses? A study comparing indoor vs. outdoor locations

Part III

DNA analysis

Chapter 5

Molecular identification of forensically important blow fly species (Diptera: Calliphoridae) from Germany

5.1 Abstract

The growth rate of blow fly larvae is highly dependent on temperature and furthermore varies between the different blow fly species infesting a corpse. It is thus crucial to identify the species collected from a crime scene correctly. To increase the quality of species identification molecular methods were applied to 53 individuals of 6 different species sampled in Bonn, Germany: *Calliphora vicina*, *Calliphora vomitoria*, *Lucilia caesar*, *Lucilia sericata*, *Lucilia illustris* and *Protophormia terraenovae*. DNA was extracted and a 229 bp fragment within the mitochondrial cytochrome oxidase subunit I (COI) was checked. The sequences of the local flies were aligned to published data of species from other countries and the application value of the analyzed region for their differentiation was studied. All species were matched correctly by a BLAST search apart from *L. caesar* and *L. illustris*. Although molecular methods are very useful especially if it is necessary to identify small fragments of insect material or very young larvae they should be used only additionally to the conventional methods. The latter are faster, cheaper and moreover are the basis for molecular species identification.

5.2 Introduction

Larval development is dependent on temperature (Bowler and Terblanche, 2008) and every species has a slightly different growth rate (Davies and Ratcliffe, 1994; Erzinclioglu, 1990; Richards et al., 2009). It is thus crucial to identify the larval species feeding from a corpse correctly to calculate the PMI_{min} properly. To ensure correct species identification, established molecular methods were transferred to the forensic field (Benecke, 1998; Sperling et al., 1994; Stevens and Wall, 1996, 1997; Wallman and Adams, 1997). Calliphoridae are one of the earliest visitors infesting a corpse with their larvae (Benecke, 2005; Lane, 1975; Putmann, 1977; Schumann, 1965). Analysis of mitochondrial DNA (mtDNA) and particulary cytochrome oxidase I gene (COI) appeared to be a useful tool in species identification among the subfamilies of Calliphoridae (Harvey et al., 2008, 2003; Wallman and Donnellan, 2001; Wells and Williams, 2005; Wells et al., 2007). MtDNA offers several advantages over nuclear DNA: the latter undergoes relatively slow mutation rates compared with mtDNA, so identification would require a much longer nucleotide sequence than is necessary with mtDNA. This enables mtDNA to provide differences in sequences between closely related species (Waugh, 2007) and therefore be useful for molecular identification purposes. Several working groups gained experience in using the method in their countries (Ames et al., 2006; Chen et al., 2004; Desmyter and Gosselin, 2009; Harvey et al., 2008, 2003; Saigusa et al., 2005; Tourle et al., 2009; Wallman and Donnellan, 2001). Surprisingly, there are no records for analysing sequences of the COI gene of blow flies originating from Germany. Even a global study and a study for European blow fly species did not include any data from Germany (Harvey et al., 2008; Vincent et al., 2000). Therefore, the aim of the study was to investigate the applicabilty of published primers and molecular methods to blow flies originating from Germany and a comparison of their COI sequences to species from several other countries.

5.3 Material and Methods

5.3.1 Specimens

53 adult blow fly specimens were captured on dead piglets exposed for insect succession experiments in the area of Bonn, Germany. Sticky traps were placed di-

rectly on the piglets. Captured specimens were removed from the trap, stored in 70
% ethanol at room temperature, and identified using morphological characteristics
(Rognes 1991). Species of three subfamilies of Calliphoridae were identified and used
for the molecular analysis: Luciliinae: *Lucilia sericata* (12), *L.ucilia caesar* (14) and
Lucilia illustris (3); Calliphorinae: *Calliphora vicina* (10) and *Calliphora vomitoria*
(6) and Chrysomyinae: *Protophormia terraenovae* (8). Sequences from specimens
collected worldwide were retrieved from the NCBI database.

5.3.2 Molecular methods

DNA was extracted from the flight muscles using the QUIAGEN DNeasy Blood &
Tissue Kit following the manufactor's instructions. A 229 bp unit within the mitochondrial cytochrome oxidase subunit I (COI) was amplified and sequenced using
the following primers: C1-J-2495 (5' CAGCTACTTTATGAGCTTAGG 3') ((Sperling et al., 1994) and C1-N-2800 (5' CATTTCAAGYTGTGTAAGCRTC) (Wells
and Sperling, 2001). For DNA amplification the HotStarTaq DNA Polymerase Kit
by QUIAGEN was used following the manufactor's instructions. Cycling conditions
were 3 min 94 °C followed by 35 cycles of: 94 °C for 60 s, 45 °C for 60 s and 72
°C for 90 s. A final extension period of 60 s at 72 °C was used, followed by holding
at 4 °C. Products were visualized using polyacrylamide gel electrophoresis followed
by silver nitrate staining. PCR products were purified using Amicon Microcon YM-
100 following the manufactor's instructions. Sequencing reactions were performed
using ABI Prism Big Dye Terminator 3.1 Sequencing Kit.

5.3.3 DNA sequence alignment and phylogenetic analysis

A BLAST search (Altschul et al., 1990) was performed on the results of the sequence
analysis of the caught flies to see if reasonable results were achieved. The most
common sequence of each of the collected species were compared to each other to
calculate the genetic differences. Additionally, the sequences of the German flies
were compared intraspecifically and to conspecific sequences originating from other
countries building a phylogenetic tree. Sequence alignment and a Neighbour Joining
tree (Saitou and Nei, 1987) were made using MEGA 4 (Tamura et al., 2007) with
pairwise deletion treatment of gaps and Tamura-3-Parameter substitution model,
bootstrap support derived from 1000 replicates and values >70 % are shown.

5.4 Results

A 229 bp fragment of the mitochondrial COI gene was successfully sequenced from
all 53 Calliphoridae specimens collected in Bonn, Germany. All specimens were
identified correctly in the BLAST searches except for *L. caesar* and *L. illustris*. In
most cases the latter were identified correctly but nearly always the second hit was *L.
caesar* and vice versa for the former. When comparing the sequences of both species
in Fig. 5.4, it became obvious how very similar they were. They diverged by only 2
bp (0.87 %). In comparison, the average intraspecific variation in *L. caesar* was 1.57
%. As the tree shows (5.4), *L. caesar* and *L. illustris* were not clearly seperated but
rather mixed within their branching. In *L. sericata* intraspecific variation was 0.5
%; for *L. illustris* it was 1.76 %. The analysed specimens of *C. vicina* varied 1.54
%, and those of *C. vomitoria* 0.17 %. The smallest intraspecific divergence value
was found in *P. terraenovae*: 0.03 %.

Analysis of interspecific divergence showed *P. terraenovae* and *C. vomitoria* to
have the least similar sequences (divergence of 29 bp, which equals 12.66 %). In
general, members of the same subfamily were closer to each other than to members
of other subfamilies. In the subfamily Luciliinae, the divergence ranged between
0.87 % and 7.4 %, within the Calliphorinae subfamily it was 4.8 %. The divergence
between the subfamilies Calliphorinae and Luciliinae was 8 - 9 %, between Luciliinae
and Chrysomyinae 9 - 11 % and between Calliphorinae and Chrysomyinae 11 - 12
%. The differences between the sequences were due only to substitutions, since no
deletions were observed.

The phylogentic tree (Fig.5.4) included not only the German specimens but also
sequences from specimens worldwide. The result was a mixed pattern of German and
worldwide specimens in the branches for each species. Thus, the German populations
were not clearly seperated from other populations, except in C. vicina. It was not
possible to distinguish between blow flies originating from Germany and those from
other countries by a BLAST search or by simple alignment.

5.5 Discussion

The aim of the study was to confirm that published molecular identification methods can also be applied on blow flies from Germany. The described method was

5. Molecular identification of forensically important blow fly species (Diptera: Calliphoridae) from Germany

```
L. sericata    ATGTAGTAGCTCACTTCCATTATGTTTTATCAATGGGAGCTGTATTTGCTATTATAGCAGGATTTGTTCACTGATATC
C. vicina      .......T..C..T...........A.....T..A.............C.................C...........C.
C. vomitoria   .......T.....T...........A.....T..A.............................A........C.
P. terraenovae .........T.....C....A.......A.......C.............T............T....TC.
L. caesar      ...............T.......A.....A...........................C.....T......
L. illustris   ...............T.......A.....A...........................G..C.....T....C.

L. sericata    CTTTATTTACAGGATTAACTTTAAATACTAAGATATTAAAAAGTCAATTTGCTATTATATTTATTGGGGTAAATTTAA
C. vicina      ......................GG...A.............A................T...A.T.
C. vomitoria   ......................CGG...A..GC.............A.................A......A.T.
P. terraenovae ..........T...C....A.......GA...T..............................T.........
L. caesar      ..C..........C.........G.A.....G....G.........A...............A..........
L. illustris   ..C..........C.........G.A.....G....G.........A...............A..........

L. sericata    CATTCTTCCCTCAACATTTCTTAGGATTAGCAGGAATACCACGACGATATTCAGACTACCCAGACGCTTACACA
C. vicina      .............................G............T........C.....T....T...........
C. vomitria    ...........................................T............T.....T............
P. terraenovae .T..........................T.......T........C..T...T.............
L. caesar      .T................T.....C.........................T...............
L. illustris   .T................T......C......................T.................
```

Figure 5.1: Sequence comparison between species originating from Bonn, Germany. Dots indicate identity to the sequence of *L. sericata*

performed successfully on 53 blow flies originating from Bonn, Germany using published primers (Sperling et al., 1994; Wells and Sperling, 2001) and extracting DNA from flight muscle tissue. Although the chosen COI sequence was rather short, it was specific enough to distinguish between the collected blow fly specimens. Harvey et al. showed likewise that the analysis of a short COI sequence is a suitable tool to distinguish between forensically important fly species in western Australia (Harvey et al., 2003). Only *L. caesar* and *L. illustris* could not be differentiated satisfactorily by comparing the chosen COI sequence. This was also described in two other studies (Vincent et al., 2000; Wells et al., 2007). These species are also very difficult to identify using morphological features (Rognes, 1991). It is therefore likely that sequences of both species provided in the database were taken from specimens that were poorly identified in the first place. Maybe in this case it would be helpful to additionally use a nuclear gene, e.g. ITS2, to provide extra information, as shown by Nelson et al. for Chrysomya species (Nelson et al., 2007). As also presented by Nelson et al., the Neighbour-Joining method is a suitable method to analyse the COI sequence to investigate whether the latter provides a sufficient resolution to identify blow flies of the genus Chrysomya (Nelson et al., 2007). The same method was used and in the resulting tree (5.4) all Calliphorinae species, apart from *L. caesar* and *L. illustris*, were resolved as monophyletic groups, despite low COI divergences between some sister species (Fig. 5.4). As to be seen, *P. terranovae* and *L. sericata* branch from the same knot indicating a closer relationship. However, the corresponding distance value was below 50 so that this branching was considered to

69

5. Molecular identification of forensically important blow fly species (Diptera: Calliphoridae) from Germany

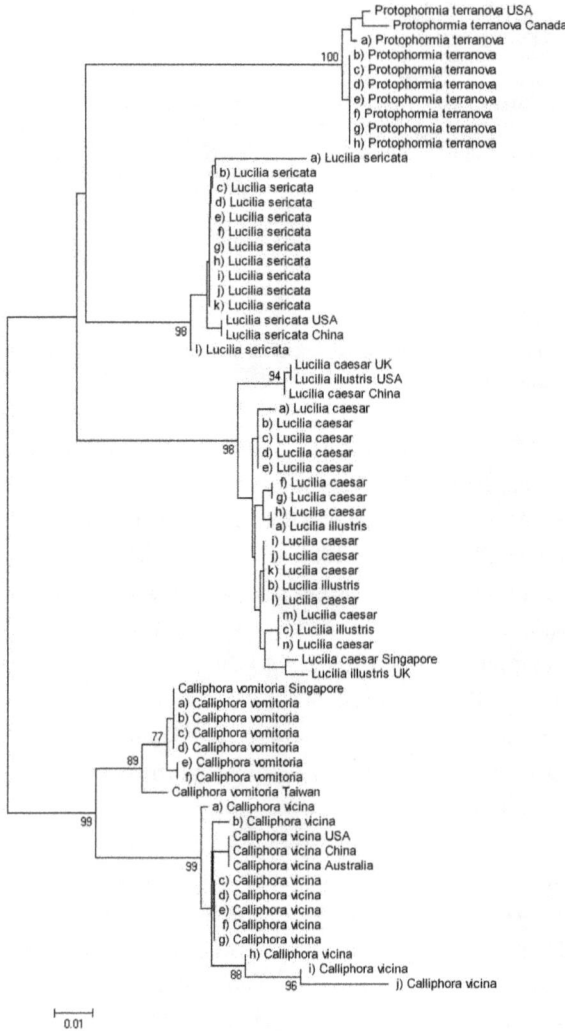

Figure 5.2: Neighbour Joining tree showing distances between species originating from Bonn, Germany consecutively numbered starting with a) (all flies were caught on dead piglets from field experiments) and species from all over the world; sequences of the latter taken from Pubmed Nucleotide Databank, their origin written behind the species name.

be unsupported. Furthermore, the interspecific divergence between Luciliinae and Chrysominae was 9 - 11 %, which also indicates an artificial relationship displayed in the tree. A closer look on *C. vomitoria* reveals that the species from Taiwan was seperated from the rest. This might be due to geographical particularities which resulted possibly in species formation. For the other german blow fly specimens investigated, identification by analysing COI was a safe way to achieve correct identification results. This can be helpful when damaged specimens or preserved larvae are to be analyzed to increase the quality of identification. It is thus not possible to decide from which country specimens of interest originated from (Harvey et al., 2008, 2003). Nevertheless, molecular methods should only be used additionally to conventional identification methods. The latter are faster, cheaper and moreover build the basis for correct molecular species identification.

5. Molecular identification of forensically important blow fly species (Diptera: Calliphoridae) from Germany

Part IV

Calculating larval age to estimate a post-mortem interval

Chapter 6

A new simulation-based model for calculating post-mortem intervals using developmental data for *Lucilia sericata* (Dipt.: Calliphoridae)

6.1 Abstract

The age of the most developed insect larvae (mostly blow fly larvae) gives reasonably reliable information about the minimum time a person has been dead. Methods such as isomegalen diagrams or calculations of accumulated degree hours (ADH) can have problems in their reliability, so in this study a new growth model was established to calculate the larval age of *Lucilia sericata* (Meigen 1826). This is based on the actual non-linear development of the blow fly and is designed to include uncertainties, e.g. for temperature values from the crime scene. Published data for the development of *L. sericata* was used to estimate non-linear functions describing the temperature dependent developmental behavior of each developmental state. For the new model it is most important to determine the progress within one developmental state as correctly as possible since this affects the accuracy of the PMI estimation by up to 75%. It was shown that PMI calculations based on one mean temperature value differ by up to 65% from PMIs based on an 12-hourly time temperature profile.

6. A new simulation-based model for calculating post-mortem intervals using developmental data for *Lucilia sericata* (Dipt.: Calliphoridae)

Differences of 2 °C in the estimation of the crime scene temperature result in a deviation in PMI calculation of 15 - 30 %.

6.2   Introduction

The basic principles of blow fly development as well as the common methods for PMI estimation were introduced in chapter one. The used methods (isomegalen diagram and ADH calculation) have limits and flaws, respectively. A new model was established to improve PMI estimations.

6.2.1   Life cycle of blow flies

Here, the life cycle of a blow fly is described in greater detail as it is essential for the following steps in establishing a new model. The development of blow flies includes four stages: egg stage, larval stage, pupal stage and imago stage (Tao, 1927). During the larval stage three instars can be separated: 1st, 2nd and 3rd instar, where the latter is divided due to behavioral changes in feeding and post-feeding larvae. Blow flies deposit egg clutches directly on the food substrate, such as a dead body (Smith and Wall, 1997), in a position where the eggs are protected and in a moist environment. This ensures a food supply for the hatching 1st instar larvae. The first three instars each undergo a moult in order to reach the next developmental stage; the stages can be distinguished by the number of respiratory slits at the posterior end of the larvae. The third instar stage lasts for longer than the first two. The larvae feed on the substrate as 3rd instars, then leave the food source to find a suitable place for pupation, entering the post-feeding stage (Arnott and Turner, 2008). About one third of the pre-adult development time is spent in the post-feeding larval stage (Greenberg, 1991). Then pupation sets in and the imago develops within the pupal case till eclosion (Lowne, 1890). This last stage persists for about half of the time of the total development.

The larval's growth rate depends on its body temperature, which is directly influenced by environmental conditions as the ambient temperatures and the heat generated by maggot aggregations (Slone and Gruner, 2007). Also, each species has its own temperature dependent growth rate.

6.2.2 Why is a new model of interest?

As described in chapter one, the method of isomegalen diagrams is only safe to use, when the temperature at the finding place of the maggots was not fluctuating. Unfortunately, this is rarely the case in outdoor scenarios. The method of ADH calculation underlies the assumption that the reciprocal of the development time (development rate) is linear over a range of temperatures. This is only the case for a distinct temperature area. In high an low temperature areas the development rate becomes non-linear. Nevertheless, it is a standard procedure to calculate the ADH values for temperature ranges in which the developmental rate is non-linear (Higley and Haskell, 2009), which results in PMIs not as accurate.

During the preparations for establishing a new model another problem was detected: a published data-set for the development of *L. sericata* (Meigen 1826) (Grassberger and Reiter, 2001) was analyzed and the corresponding ADH values for these data were calculated. Fig. 6.1 shows the calculated ADH values for a base temperature of $\Theta_0 = 8°C$ (as calculated by a linear regression analysis for the used data-set). In the figure a new effect can be observed: for the younger and also shorter developmental phases the ADH values are nearly constant over the complete range of temperatures, but for the post-feeding and the pupal stages the ADH values are strongly temperature dependent.

Furthermore, it is highly problematic that uncertainties for temperature measurements from a crime scene cannot be taken into account in either of the commonly used methods for PMI estimation. It is difficult to determine the actual temperature controlling the larvae at a real crime scene. Since temperature is the variable that most influences development, it is crucial to consider it as accurately as possible. The standard procedure is to use temperatures of the nearest weather station for the desired time frame and correct them by applying a regression starting from temperatures measured at the crime scene, when taking the larvae as evidence (Archer, 2004a). The corrected values still contain uncertainties that cannot be accounted for by the methods currently used for PMI determination. No information exists for either model about the quality of the method or the error intervals of the calculated PMIs.

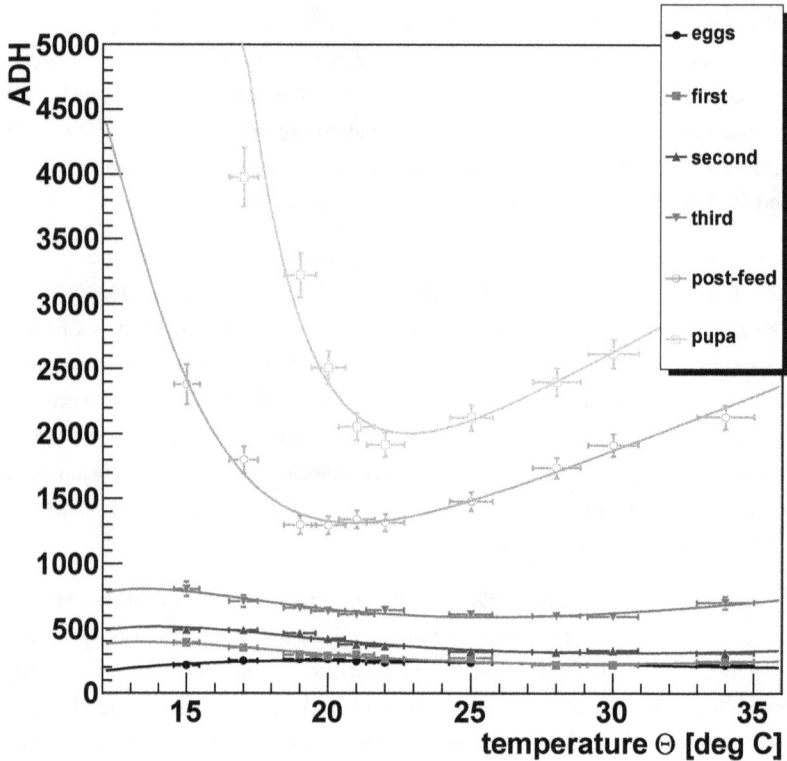

Figure 6.1: Calculated ADH values for the development of *L. sericata* using eq. (1.2)
(data points) and fitted functions (lines) calculated using eq. (6.1) and estimated
parametres (Table 6.1), data by Grassberger and Reiter 2001.

6.3   New approach for PMI determination

Developmental data for *L. sericata* generated under different temperatures was analyzed and an individual exponential function for each developmental stage was fitted
after applying a non-linear regression analysis of the data. Data used as input to the
model were published by Grassberger and Reiter (2001) and represent the minimal
time in hours to complete each larval phase (egg stage = stage 0, 1st instar = stage
1, 2nd instar = stage 2, 3rd instar feeding = stage 3, 3rd instar post-feeding = stage
4 and pupal stage = stage 5) until eclosion of the adult blow fly. The used data set
is one of the rare sets which covers a lot of temperatures and the resulting growth
curve seems to represent growth behavior well (see original paper). Unfortunately,
Grassberger and Reiter do not give any error values for their measurements, so an
error for the developmental times of about 1 hour was assumed. These authors used
250 g of raw beef liver in plastic jars, and placed 100 eggs on the food substrate.
The jars were placed in a precision incubator. At each temperature regime the procedure was repeated 10 times. Every 4 hours, four of the most developed maggots
were removed from the plastic jars, killed in boiling water, and preserved in alcohol
(Adams and Hall, 2003) and their stage of development was determined.

6.3.1   Data fit

The new larval growth model is based on the data shown in Fig. 6.2, in which the
duration of each developmental stage was measured as a function of temperature
(Grassberger and Reiter, 2001). These data points were fitted with an exponential
function of the form:

$$T_\alpha(\Theta) = a_\alpha \cdot \exp\left(-\tau_\alpha \cdot \Theta\right) + T_{0,\alpha} \tag{6.1}$$

where T_α is the duration of one developmental stage α as a function of temperature Θ. The parameters fitted for the different stages are shown in Table 6.1. The
parameter τ_α defines how strongly the time interval depends on temperature; the
higher the parameter in Table 6.1, the steeper is the gradient of the fitted curve. $T_{0,\alpha}$
represents the minimum time interval required for finishing a certain developmental
stage and a_α provides the absolute normalization. The developmental stages of the
maggots were determined every $\Delta T = 4$ h, such that time measurement errors are set
to $\sigma_T = \Delta T/\sqrt{12}$ following an uniform distribution. It is assumed that the maggot

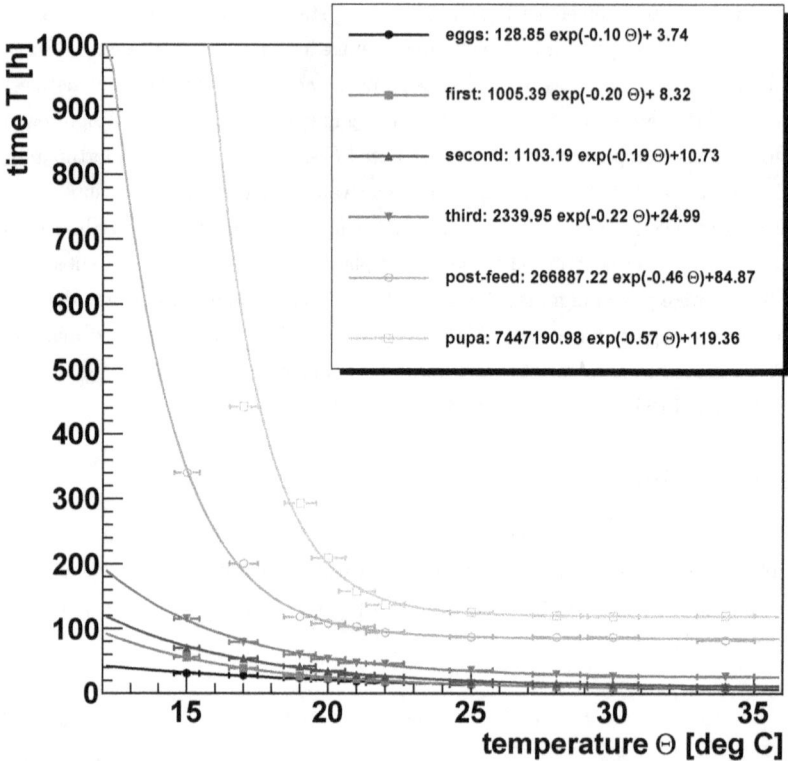

Figure 6.2: Developmental data of *L. sericata* with fitted functions (eq. (6.1)), data
by Grassberger and Reiter 2001.

6. A new simulation-based model for calculating post-mortem intervals using developmental data for *Lucilia sericata* (Dipt.: Calliphoridae)

Table 6.1: Fitted parameters for the development-time-function $T_\alpha(\Theta)$.

α	Stage	a_α [h]	τ_α [°C^{-1}]	$T_{0,\alpha}$ [h]
0	eggs	$1.28 \cdot 10^2$	0.10	3.74
1	1st	$1.00 \cdot 10^3$	0.20	8.32
2	2nd	$1.10 \cdot 10^3$	0.19	10.73
3	3rd	$2.34 \cdot 10^3$	0.22	24.99
4	post-feed	$2.67 \cdot 10^5$	0.46	84.87
5	pupa	$7.45 \cdot 10^6$	0.57	119.36

body temperature is known to an accuracy of 3 % in order to take into account uncertainties about differences between ambient and maggot body temperature. The parameters a_α, τ_α and $T_{0,\alpha}$ were determined by minimizing the sum of error squares. As seen in Fig. 6.2, the exponential function accurately models the behavior during all developmental stages and will be used below. In all stages the developmental duration at temperatures below 24 °C starts to rise exponentially. Fig. 6.1 shows the calculated ADH values corresponding to eq. (1.2) (data points in figure). In addition, the figure shows the function $\text{ADH}_\alpha(\Theta) = T_\alpha(\Theta) \cdot (\Theta - \Theta_0)$. $T_\alpha(\Theta)$ is calculated by eq. (6.1) with the previously fitted parameters (Table 6.1). Again, the functions give reasonable description of the data. Nevertheless, the model is an empirical one, based on the observations of the data points generated by Grassberger and Reiter (2001).

6.3.2 PMI calculation

For European and especially German temperatures, calculation of the total developmental duration must allow for non-linear temperature behavior in order to ensure accuracy. The basic idea underlying a new approach in PMI determination is to follow an ambient time-temperature profile $\Theta(t)$ backwards in time starting from the time point t_F at which the maggots of interest were collected. The idea of backwards calculation is obviously similar to the ADH method, but in the new model the important improvement is the calculation method for determining the progress of larval development. The latter is calculated successively during certain time steps using the fitted functions corresponding to the current developmental stage introduced in Fig. 6.2. In each stage α the relative developmental progress is P_α (values 0 - 1) where 0 is the starting point and 1 is the end point of each developmental

81

6. A new simulation-based model for calculating post-mortem intervals using developmental data for *Lucilia sericata* (Dipt.: Calliphoridae)

stage; e.g. a maggot in the middle of the post-feeding stage is $P_4 = 0.5$, at the end of the post-feeding stage it is $P_4 = 0.9$ and so forth. The developmental duration $t_{\alpha,0}$ spent in each individual stage is calculated by solving the relation:

$$P_\alpha = \int_{t_{\alpha+1,0}}^{t_{\alpha,0}} \frac{dt}{T_\alpha(\Theta(t_F - t))}. \tag{6.2}$$

where $dt/T(\Theta(t))$ is the infinitesimal relative development.

The calculation starts with the developmental stage of the maggot at the time of collection, summing the developmental progress backwards until the beginning of the egg stage is reached. The calculation for each collection stage uses $t_{\alpha+1,0} = t_F$. The total development time t_0 or post-mortem interval (PMI) is then given by

$$t_0 = \sum_\alpha t_{\alpha,0}. \tag{6.3}$$

For the new model a program was written in C++ using Root (http://root.cern.ch/) as analysis software. This program includes all mentioned mathematical steps and produces the figures shown here as output. For each new PMI calculation the corresponding temperature profile can be inserted and individually chosen uncertainties can be included.

6.3.3   Consideration of uncertainties by Monte-Carlo simulation

To explore the uncertainties in the total developmental duration, a Monte-Carlo simulation was applied, which is commonly used for simulations in life sciences (Mansson et al., 2005). It is a method for calculating one final uncertainty after considering all statistically independent uncertainties that influence for example the larval age. The mean PMI with corresponding standard variation is calculated n times taking into account and varying all uncertainties described in the following. First, the developmental profiles $T_\alpha(\Theta)$ have uncertainties due to the measurement procedure. Second, the time-temperature profile from the collection scene is not known precisely and must be approximated using temperature values from nearby weather stations. The variations are introduced for each model as follows:

Development profile: The mean duration values of the temperature-time data

82

are randomly smeared with a uniform distribution with corresponding error σ_T; for the maggot body temperature $\sigma_{\Theta,b}$ a Gaussian distribution is used. New fits with the function in eq. (6.1) are performed for each stage.

Time-temperature profile: Deviations between the temperature profile at the collection scene and the nearest weather station are accounted for by Gaussian smearing of time t and temperature Θ, with the corresponding errors σ_t and σ_Θ as width for each data point. σ_Θ can be inserted in the model's calculation individually, dependent on the differences between the temperatures at finding place and weather station.

The PMI for a mock crime scene was calculated using the following parameters: $\sigma_T = 4/\sqrt{12}\,\text{h}$, $\sigma_t = 1\,\text{h}$, $\sigma_\Theta = 2°\text{C}$, $\sigma_{\Theta,b} = 3\,\%$ for 10000 models for a fixed collection stage progress of $P_\alpha = 0.5$. The results are shown in Fig. 6.3 *(upper part)* which is a direct output of the new program that calculates the PMI. The lower time axis defines the progress of the temperature profile forward in time, representing the time frame of interest. The temperature profile used here (black line) is taken from the minimum and maximum temperatures in May and June 2008 measured at Cologne/Bonn airport. The right end of the diagram marks a fictional time point of maggot collection and therefore the starting point for PMI calculation. The upper time axis depicts the PMI backwards in time starting from the moment of maggot collection. For each developmental stage the PMI was calculated by following a linear interpolation between the maximum and minimum temperatures. The resulting histograms illustrate the PMI distribution for each stage and show a clear single peak structure. The arrows on the top show the 1-standard deviation interval for each stage around the mean PMI value, and range between 0.1 and 1.2 days (depending on the stage). Since no data points below temperatures $\Theta < 15°\text{C}$ were measured, the functions $T_\alpha(\Theta)$ were extrapolated to lower temperatures. As expected, the PMI and the corresponding standard deviation increase with higher developmental duration (see arrows above histogram).

Since the exact progress within the developmental stage at collection time is most of the time also unknown, a third uncertainty is introduced:

Stage progress: The developmental stage at collection time was determined only to integer precision, so that it is assumed the exact progress is an uniformly distributed value between 0 and 1. Consequently, the starting value for the

PMI calculation P_α at time t_F is randomly and uniformly chosen within the interval [0,1] for each model.

Fig. 6.3 *(lower part)* shows the PMI calculation for the same parameters as before, but without setting the progress of the development for each stage to a fixed value. The 1-standard deviation values increase by 0.3 to 3.3 days. The resulting uncertainty in the progress of the stage contributes about 75 % to the total PMI error interval. A similar result was also found by Tarone (Tarone and Foran, 2008) using generalized additive models for the analysis of the growth rate of *L. sericata*: the analysis of the stage alone achieved better results than the analysis of the length or weight alone. In addition, the histograms show deviations from a clear single peak structure, e.g. for the pupal stage, implying that the probabilities for a PMI of 21 days and 26 days are nearly the same. To use the new model, the crucial parameter is therefore the correct determination of the progress in each developmental stage.

6.3.4 Estimation of temperature at the location of maggot collection

The impact of correct temperature determination at the maggot collection scene is shown in Fig. 6.4. The data points represent the mean PMIs with an error bar of 1 standard deviation as a function of collection stage for three different temperature profiles. The triangles show the PMIs for the original temperature profile as measured at Cologne/Bonn airport. The bullets (squares) show the results for the same profile but subtracted (and added) by 2 °C. As expected, the PMIs and the corresponding standard deviations of the lower (higher) temperature profile increase (decrease) relative to the nominal profile. These differences in temperature of 2 °C give rise to an effect of 15 - 30 %. That implies that a miscalculation of the temperature at the crime scene of 2 °C will result in a miscalculation of the PMI by 15-30 %. The later the stage, the greater the deviation from the actual PMI.

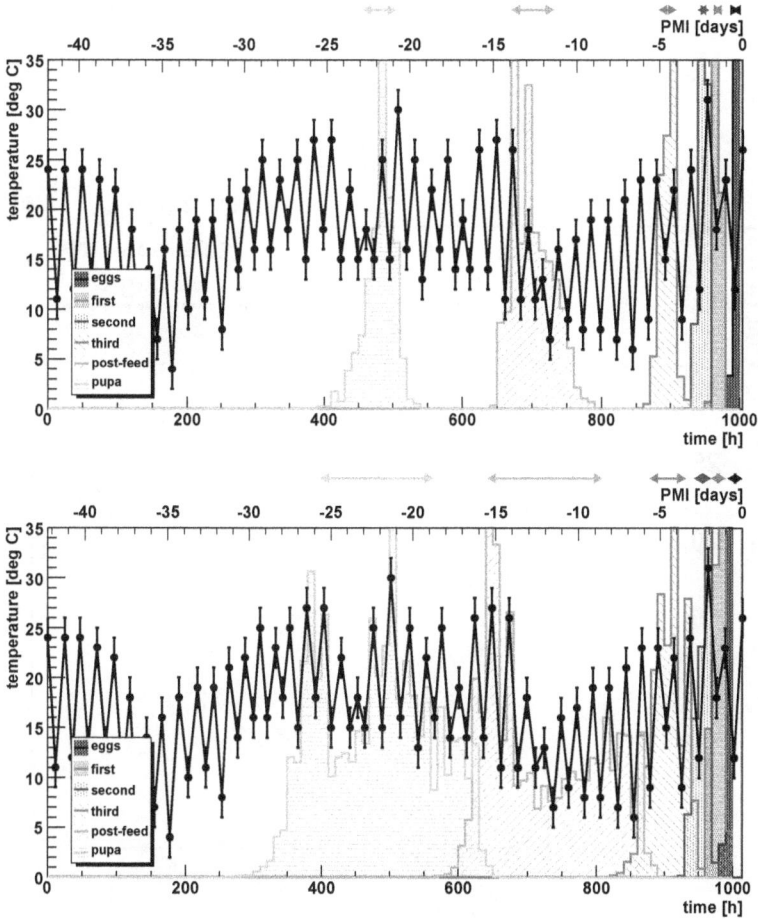

Figure 6.3: Excerpt of the temperature profile during May and June 2008 in
Cologne/Bonn, Germany (www.wetteronline.de). The histograms illustrate the PMI
distributions of the different random models for a certain collection stage α. The
arrows at top show 1-standard deviation intervals around the mean PMIs for each
stage. *Upper part:* PMI calculation with parameters $\sigma_T = 4/\sqrt{12}\,\mathrm{h}$, $\sigma_t = 1\,\mathrm{h}$,
$\sigma_\Theta = 2°\mathrm{C}$, $\sigma_{\Theta,b} = 3\,\%$ for 10000 models for a fixed collection stage progress of
$P_\alpha = 0.5$. *Lower part:* PMI calculation as before but the progression of the stage
is not fixed.

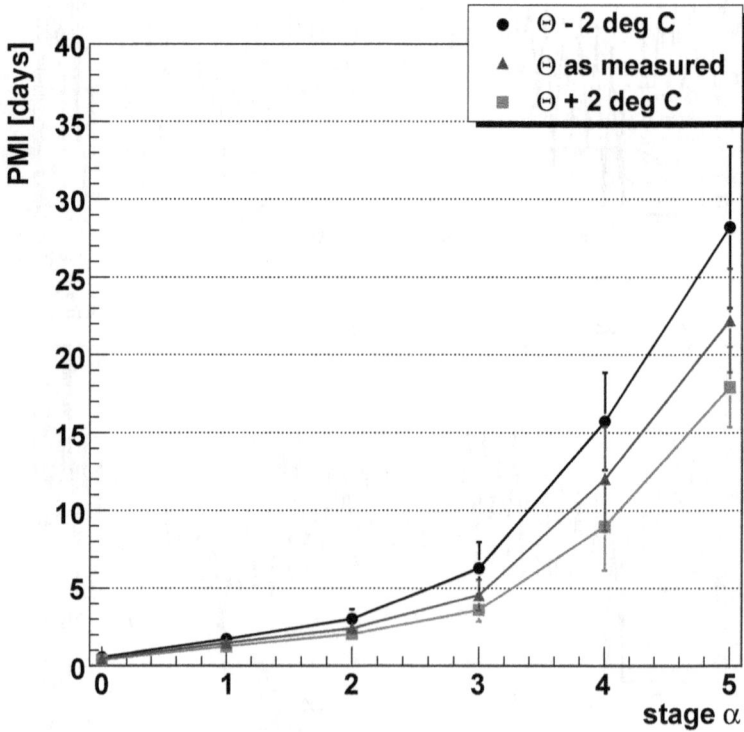

Figure 6.4: PMI vs. collection stage for May/June 2008 Cologne/Bonn temperature
profile in comparison to profiles ± 2 °C.

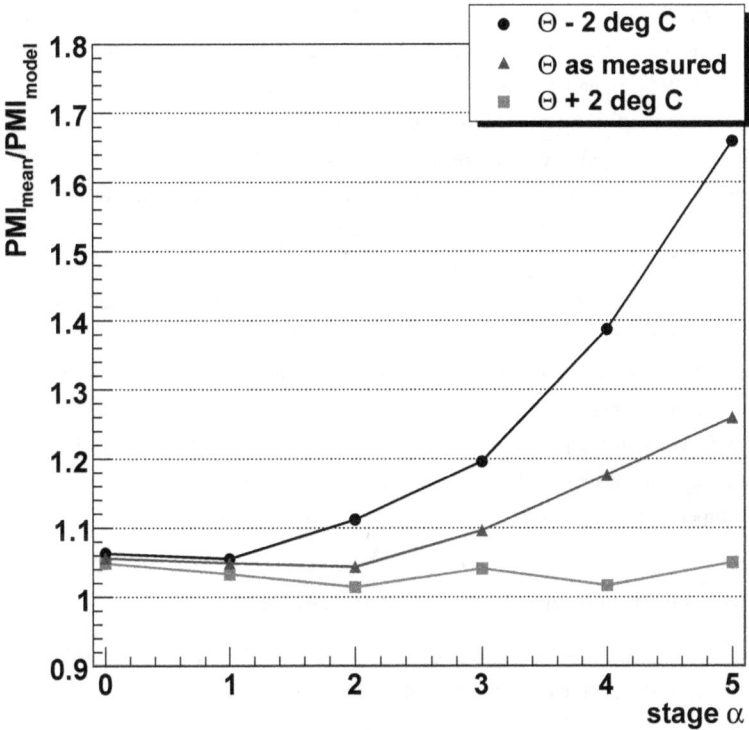

Figure 6.5: Calculation of PMI using mean temperatures and the actual temperature profile for May/June 2008 Cologne/Bonn. Additionally, the temperature profile is shown ± 2 °C.

6.3.5 Comparing PMIs based on a mean temperature and a 12-hourly temperature profile

In Fig. 6.5, PMIs calculated using the three temperature profiles introduced previously are compared to PMIs calculated using mean temperature values corresponding to the originally used temperature profiles in the temperature interval $[t_0, t_F]$. The calculated mean temperatures were as follows (calculated for the time frame till completion of each stage): stage 0 = 18°C, stage 1 = 19 °C, stage 2 = 20 °C, stage 3 = 19 °C, stage 4 = 16 °C, stage 5 = 17 °C. The PMI values based on the temperature profile and those based on a mean temperature value agree to within about 5 % for the high temperature value (original profile + 2 °C) in all stages. The deviation between the mean temperature profile and the original temperature profile exceeds the 10 % level starting at the 3rd instar feeding stage, and increases to 25 % in the pupal stage. This effect becomes even more dramatic for the low temperature profile (original profile subtracted by 2 °C). Starting from the 2nd instar stage, the deviation increases from about 10 % up to about 65 % for the pupal stage. This means that use of mean temperature values overestimates the influence of low temperatures and underestimates periods of high temperatures. The effect should be larger if the mean temperature during the development is lower still, e.g. in spring or fall. In general, more data points are needed for the developmental duration at low temperature ranges to provide more reliable statements.

6.3.6 Does the model work in a real case?

The PMI for a real case was calculated where the actual PMI was known due to a confession of the offender. At the end of August 2007 the victim was killed in early morning and was found 4 days later also in the morning on a grassland. This leads to a PMI of approximately 96 hours. The victim was stabbed to death and had several wounds which would act as attractant to the blow flies. It can be assumed that blow flies started ovipositing early after death occurred (Reibe and Madea, 2010a). Autopsy was performed directly after the corpse was recovered and several 2nd instar larvae of *L. sericata* were collected. The largest larvae measured 6.1 mm. Hourly temperature values were taken from a weather station 10 km away. The mean temperature was 16 °C. Using Grassberger and Reiter's isomegalen diagram for a larva measuring 6 mm and a mean temperature of 16 °C results in a time interval of 3.2 days plus 30 hours (larval development time plus egg period). In total, a PMI of 107 hours is indicated. This would shift the time of oviposition to nighttime, which is a highly unlikely event (Amendt et al., 2008). The same data can be used to calculate the ADH value for *L. sericata* for reaching 6 mm to calculate the PMI not based on the mean temperature but on hourly data. As mentioned earlier, a regression analysis of the data set reveals a base temperature of 8 °C. The corresponding ADH value is therefore 856, based on the equation:

$$ADH = 107(16 - 8) \tag{6.4}$$

Subtracting the hourly ADH values, estimated by the temperature values from the weather station and the base temperature, from the starting value of 856 results in a PMI of 101 hours.

For usage of the new model information about the progress of the 2nd instar larval stage was required. In the original work of Grassberger and Reiter (2001, Figure 1) a figure is included showing the temperature dependent growth of the larvae and also in the diagram the time points of each moult are shown. According to this figure, the 2nd instar stage sets in after the larvae have reached a size of approximately 4 mm and ends when the larvae have reached a size of approximately 8mm. As the largest larvae collected measured 6 mm, P=0.5 was chosen as progress for the larval stage. The hourly temperature profile was inserted and a temperature error of 1 °C was chosen. The result of the calculation was a PMI of 99 hours (SD = 3 hours).

The calculations of a PMI in a real case show that all three methods give reasonable results. Furthermore, it becomes obvious that the new model is a possible alternative for the existing methods with the benefit of directly providing a standard deviation for the calculation.

6.4   Conclusion and outlook

The new model improves the larval age calculation in specific ways. It can be used in non-linear parts of the temperature dependent development, and includes individually defined uncertainties for a temperature profile determined retrospectively from the nearest weather station. In the new model the temperature profile plus the determination of the larval stage are translated into a mean PMI as well as a standard deviation. PMI calculation using mean temperatures, however, can lead to severe deviations from the real PMI.

So far, the main uncertainty arises from the fact that the developmental stage is determined only on a 1 - 6 scale (egg, 1st instar, 2nd instar, 3rd instar feeding, 3rd instar post-feeding and pupae). As shown above, 75 % of the uncertainties in the model depend on the exact determination of the developmental progress, and additional length values, as shown for the PMI calculation in the real case, will propably increase its accuracy leading to more accurate PMI calculations. Moreover, the next step is to produce own growth data with known error values to refine the inclusion of uncertainties that are only rough estimates at the present time and to improve the till now only empirical model.

Nevertheless, the new PMI calculation program is suitable for use in forensic case work as a general tool for PMI determination. Scientists from every country or climatic region can incorporate their own growth values for different species and ensure a high accuracy in PMI determination.

Chapter 7

Growth modeling of *Calliphora vicina*

7.1 Introduction

In the previous chapter, a new method to estimate the larval age was introduced. The data used for establishing the new model are published values for the development of *Lucilia sericata* (Grassberger and Reiter, 2001) from Vienna. Sarah Gulinski repeated those experiments using flies originating from Bonn for her Diploma thesis in 2009 (unpublished data). The aim of her thesis was to compare both sets of developmental data, fitting the data from Bonn in the new model and to validate the new model. The comparison of the developmental data from Bonn and Vienna revealed that the larvae from Bonn needed more time to finish development at all temperatures.

For the thesis, only five constant temperatures could be applied in contrast to ten temperatures for the data collection from Vienna. The temperature values were 13, 19, 20, 21 and 25 °C. The three middle temperature values were chosen because in the data of Grassberger and Reiter big differences were stated for the total development time. Under constant temperatures of 20 °C total development was 4.7 days shorter than under constant 19 °C , between 20 °C and 21 °C the difference was 3 days for completing development. In the experimental setup for data collection for blow flies from Bonn it was tested if these results could be reproduced. The resulting developmental curves for both data sets are shown in Figure 7.1 A (Grassberger and Reiter data) and C (Gulinski data). The data from Grassberger and Reiter

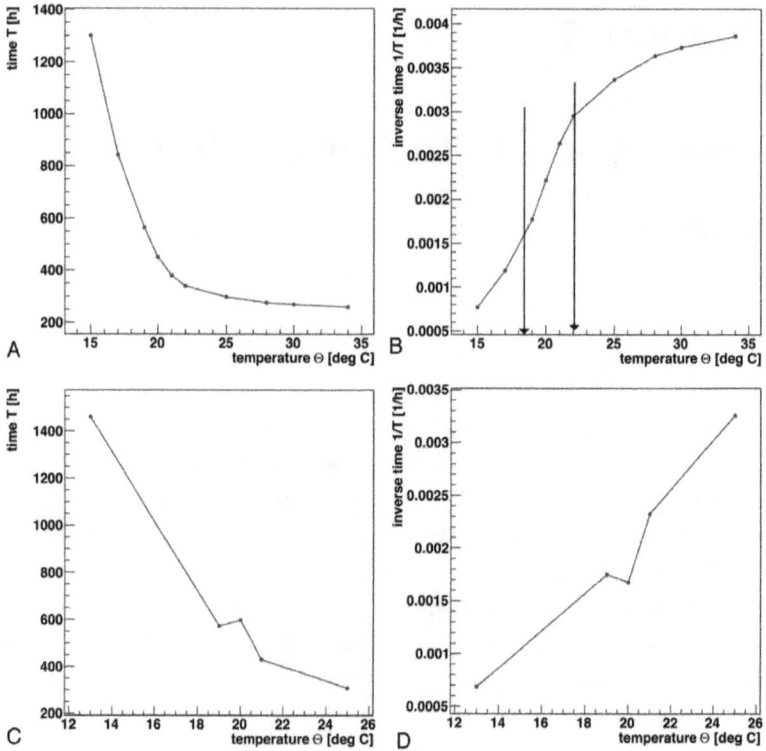

Figure 7.1: Development of *L. sericata*. A: Time in hours for completing development, B: developmental rate [1/devtime], area between arrows: temperature range where the developmental rate can be regarded as linear (A and B Grassberger and Reiter data, n= 10 for each temperature). C: Time in hours for completing development, D: developmental rate [1/devtime] (C and D Gulinski data, n=4 for each temperature).

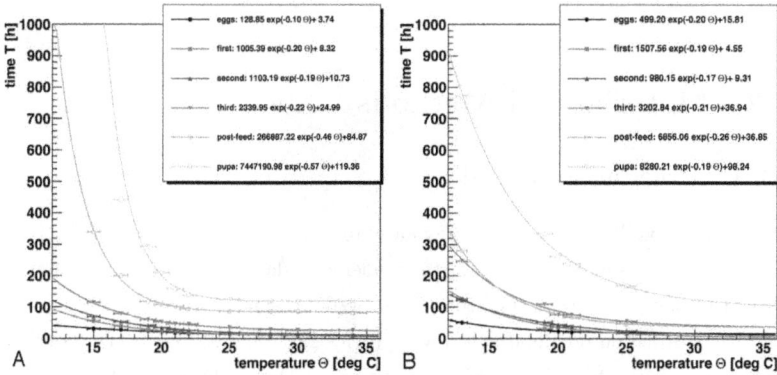

Figure 7.2: Development of *L. sericata*. A: Time in hours for completing each developmental stage of *L. sericata*, A: Grassberger and Reiter data, B: Gulinski data.

display the same curve shape proposed for poikilothermically development (Janisch, 1928), as discussed in chapter 1. Moreover, the corresponding developmental rate reveals the supposedly linear part in which the ADH method can be applied (Fig. 7.1 B, area between arrows). According to this figure, the temperature range is between 18.5 and 22 °C. The developmental curve (Gulinski data) based on only five temperatures does not resemble the theoretically proposed curve shape (Fig. 7.1 B), as intermediate values between 13 °C and 19 °C and also temperatures above 25 °C are missing. Those would probably adjust the curve to the correct form. Furthermore, the resulting developmental rate could be interpreted as linear over the complete range of temperatures. These data would therefore be a poor start for developing a new model. However, implementing these data into the existing model gave good results. The model uses the form of the functions fitted to the data of Grassberger and Reiter (Fig. 7.2 A) but bases the calculation on whichever developmental data is put in (Fig. 7.2 B).

To check the model's universal character concerning other species, developmental data were produced for the blow fly *Calliphora vicina* originating from Bonn and implemented in the model. The chosen temperatures were the same as for data collection of *L. sericata* data. Again, the basis of the model are the fitted functions based on the Grassberger and Reiter data but the values established for *C. vicina*

93

are used to calculate the larval age.

7.2 Material and Methods

7.2.1 Breeding of the flies

As originally published data from Vienna were inserted in the new model for PMI calculation, it was important to test the model with data established for the development of blow flies originating from Bonn. Wild specimens of *L. sericata* and *C. vicina* oviposited on dog food which was provided in the open field. The hatching larvae were reared till eclosion of the adult flies. These adults were the starting population for the breeding. To capture genetic variation in the breeding specimens from real cases were mixed with adults from the starting population.

Adult blow flies of *C. vicina* were kept in cages made of a wooden frame covered with gauze (60x60x60 cm) (Fig. 7.3 A). 50 - 100 specimens were kept in the cages. Water and sugar was provided constantly and pieces of wet dog food (Orlando, Lidl) were served as protein source and oviposition site (Fig. 7.3 B). The room had a window and in addition to daylight two light sources were present (Reptiglow 2.0) in a 16:8 light:darkness (L:D) cycle to prevent larvae undergoing diapause.

In the morning of a new experimental run, a small sample of dog food (30 g) was provided for the flies after a period of 1-2 days without access to meat. The dog food was checked every 30 minutes. When egg patches were present, the dog food was removed from the cage and prepared for the climate chamber as follows. A plastic box (30x20x20 cm) was inlaid with paper towels to provide a suitable pupation site for the larvae, a second, smaller plastic box (20x10x10 cm) was filled with 500 g of dog food and put inside the larger plastic box. On the large amount of meat the small pieces of meat (with egg batches) were added. The box was closed with a lid that had several ventilation holes. Between box and lid a thin layer of gauze was stretched to prevent escape of wandering larvae. The box was labeled properly and put in the climate chamber (KB 710, Binder). Five different constant temperatures were applied to at least two boxes prepared as described previously: 13, 19, 20, 21 and 25 °C. The climate chamber was also equipped with a light source in a 16:8 L:D cycle. The two boxes for each of the temperature regimes were put in the chamber in 5 to 8 hour intervals. In the morning of the second day, the boxes were

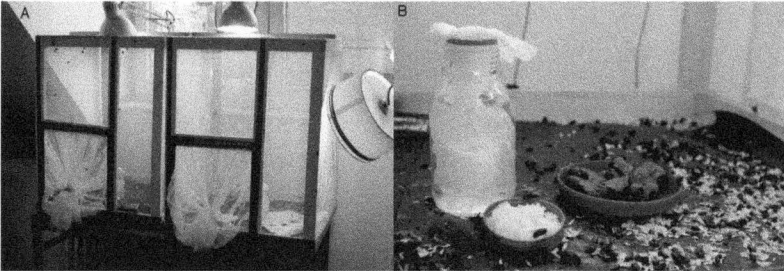

Figure 7.3: A: Cages for adult blow flies, B: Nutrition and egg deposition site for the blow flies.

checked for hatched larvae. In hourly intervals the boxes were checked and the time of hatching was noted. The following days 4-6 larvae were taken from the box twice a day (10:00 am and 6:00 pm), killed in boiling water and preserved in 70% EtOH. The age in hours, the length [mm] and the stage of the larvae were determined. The post-feeding phase was defined as the onset of wandering when the larvae left the food and were crawling underneath the paper towels. For implementing the data in the model, a table with the mean durations in hours for finishing each stage in relation to the temperature was generated.

7.2.2 Validation

To use the generated data for the development of *C. vicina* in the new model and check for their reliability, different methods were applied. Firstly, *C. vicina* oviposited on dead piglets (exposure time of the piglets about two hours) which afterwards were put in boxes similar to those used for the breeding experiments. The boxes with the dead piglets and freshly laid egg patches were put on the roof of the Institute of Forensic Medicine. Next to the boxes a data logger (Ebro Ebi-6) was placed to record the temperature every 30 minutes. The boxes were checked daily and the onset of wandering and pupation of the larvae as well as hatching of the adult flies were noted. The temperature data were used to calculate the PMI starting from the days of hatching or alternatively the collected larval stage. The PMI was calculated using the new model and the ADH method and the results were compared to the real PMI. As reference data for the ADH values, published data were used (Ames and Turner, 2003). All calculations using the new model were

95

modeled 100 times and the base temperature of development, which is considered
by the model, was set to 4 °C (Ames and Turner, 2003). Furthermore, the PMI
was calculated with both methods using temperature data from the nearest weather
station (Institute of Meteorology, University of Bonn) as well as temperatures calculated by a regression formula based on 24 hours temperature data recorded by the
data logger and temperature data from the weather station. Secondly, the climate
chamber was set to fluctuating temperatures (15/25 °C) and larvae were reared as
described previously. Again, the developmental stages were noted and the PMI
was calculated using the new model as well as the ADH method. The results were
compared to the real PMI. Thirdly, one real case was evaluated where the missing
interval and the post-mortem interval have most likely been identical so that the
results of the PMI calculation of the different methods could be compared to the
missing interval.

7.3 Results

7.3.1 Development of *C. vicina* under constant tempera-
tures

Five constant temperatures (13, 19, 20, 21, 25 °C) were applied to larvae of *C. vicina*. The minimum duration for finishing each stage was determined and is shown
in Table 7.1. The total developmental time and the developmental rate are shown
in Fig. 7.4. Obviously, the total developmental time decreases with increasing temperatures. However, the curve shape does not resemble the theoretically proposed
shape for poikilothermically development. As only five data points are available, it
is unclear how further data points would be distributed in between the existing data
points. Nevertheless, as pointed out before for the data for *L. sericata*, the data
for *C. vicina* are implemented in the model which is based on the ten data points
determined by Grassberger and Reiter. The model fits an exponential function for
the later PMI calculation to each new dataset. The resulting functions for the data
of *C. vicina* are shown in Figure 7.6.

As discussed in the previous chapter, for the best PMI results calculated by the
model it is important to estimate the progress of the developmental stage as precisely
as possible. In Figure 7.5 the distribution of the length of larvae in the 2nd and

7. Growth modeling of *Calliphora vicina*

Table 7.1: Minimum duration for finishing each developmental stage in hours of larvae of *C. vicina*. SD = Standard deviation, pf= post-feeding larvae, pp = pupae

°C	egg	SD	1st	SD	2nd	SD	3rd	SD	pf	SD	pp	SD	total	SD
13	48	1.4	69	9.0	73	2.1	218	10.1	72	5.7	602	6.1	1082	7.1
19	28	1.5	43	4.2	21	1.5	148	9.2	48	4.2	336	4.2	624	12.9
20	24	1.4	41	4.9	27	1.4	200	14.1	72	4.0	300	7.1	564	7.3
21	22	1.4	42	8.1	31	3.0	113	13.7	32	14.4	300	20.0	540	32.8
25	20	0.7	28	4.2	19	2.9	91	4.5	82	12.1	168	5.3	408	8.5

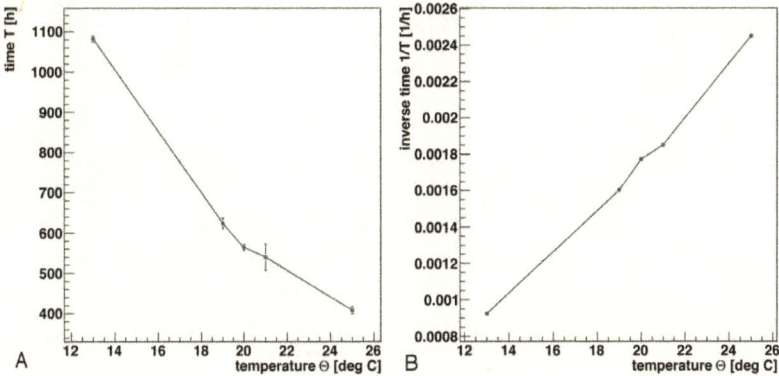

Figure 7.4: Development of *C. vicina*. A: Time in hours till eclosion, B: developmental rate.

3rd instar stage is shown for constant temperatures of 19, 20, 21 and 25 °C. The growth of *C. vicina* in the 2nd instar stage seems to be linear but in the 3rd stage it is not. The non-linear behavior of the development of larvae in the 3rd instar impedes the use of length as an indicator to estimate the progress of development. As shown in Figure 7.5, a larva of 17 mm can be found throughout two thirds of the 3rd instar stage.

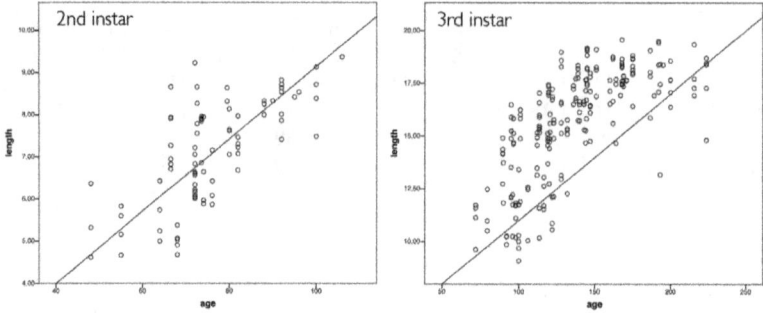

Figure 7.5: Correlation of the length of 2nd and 3rd instar larvae to their age in hours, data produced at 19, 20, 21, 25 °C and pooled for the plot.

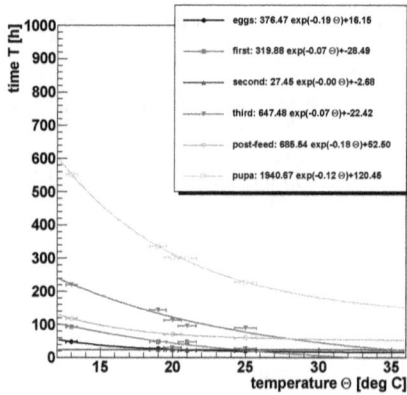

Figure 7.6: Duration of development of each stage of *C. vicina* with fitted function as determined by the model.

98

7.3.2 Validation of the data for *C. vicina* implemented in the new model

Piglets

For the validation of the model using the developmental data established for *C. vicina* under five constant temperatures three dead piglets were exposed on the roof of the Institute of Forensic Medicine, Bonn at different times. The egg batches were deposited by *C. vicina* on the piglets inside the breeding cages to assure that the eggs are deposited by the correct species and also to observe the time of oviposition. The first piglet was exposed in July 2009. The total developmental time from egg till eclosion of the adults were 21 days (509 hours). The mean temperature was 21.5 °C . The actual temperature was recorded every 30 minutes and was used for the calculations. The model calculated 517 hours (SD = 15.2) till the end of the pupal stage after the temperature error was set to 0.1°C as the data logger was lying directly next to the box. The real PMI lies within the given standard deviation. The ADH method calculated 498 hours. As mentioned before, the method can not give a standard deviation. Using hourly temperature values from the meteorological station of the University Bonn the model calculated a PMI of 547 hours (SD = 7.2, temperature error = 1 °C), which is much too long. The mean temperature from the weather station was 20.2 °C . The ADH calculation based on the hourly temperature values measured at the weather station resulted in 533 hours post-mortem. Using a regression formula (see eq. 7.1) to adjust the temperature values from the weather station (x) to the actual values recorded on the roof the model calculated a PMI of 528.2 hours (SD = 9.9, temperature error = 1 °C). The ADH method calculated 496 hours post-mortem. The following regression formula was used:

$$Temperature = -2.563 + (1.163 * x) \qquad (7.1)$$

Table 7.2 shows the percentage deviation from the real PMI after applying both methods (new model and ADH method) using three different temperature datasets: recorded by the data logger (l), measured at the weather station (w) and values calculated using a regression formula (r). The smallest deviation was achieved when using the new model and the actual recorded temperature (1.6 %). With both methods the deviations were largest when using temperature data from the weather station. The regression-based temperature values achieved better results than the data from the weather station for both methods. Using the actual temperature data and the regression-based data in the ADH method underestimated the PMI. However, using the new model for the calculation slightly overestimated the PMI.

The second piglet was exposed at the end of July 2009. For this trial, larval age was calculated for 3rd instar larvae. All later stages could not be included as heavy rainfall influenced the development drastically, as the larvae either drowned or eclosion was severely delayed. The progress of the developmental stage was set to random, the temperature error to 0.1 °C. The larvae were collected after 130 hours of development and the mean temperature was 21.9 °C. The program calculated a PMI of 135.8 hours (SD = 29.1). As the stage progress was set to random, the SD is rather large. The ADH calculation could only be conducted using the literature value for finishing the 3rd instar stage which resulted in 172 hours. However, as the larvae had not reached the end of the 3rd instar stage at collection time this value is only a very rough estimation. In 1984, Reiter published developmental data for *C. vicina* reared at 22-23 °C. He shows a table including the age of the larvae in days and the corresponding length. The larvae collected from Pig 2 measured in average 18.5 mm. According to the Reiter table, the larval age of the collected larvae ranged between 97.4 and 133.4 hours (Reiter, 1984). The real PMI lies within this range but each value within this range is as probable as any other value within the range.

The model calculation based on temperatures from the meteorological station resulted in 152.34 (SD = 39.8) when the stage progress was set to random. The mean temperature was 19.9 °C. The regression formula to adjust the data from the weather station (x) was as follows:

$$\text{Temperature} = -3.082 + (1.25 * x) \tag{7.2}$$

After applying the regression analysis to the data from the weather station the model calculation resulted in a PMI of 138.5 hours (SD = 38.9) when assuming a temperature error of 1 °C. In table 7.2 the percentage deviation is shown only for the calculation of the PMI by the new model and the different temperature sets. The deviation using the actual temperature values was 4.5 %. Again, using the temperature values based on the regression formula achieved better results than using the data from the weather-station (station: 17.2 % , regression: 6.5 % deviation). In general, as discussed before, setting the progress of the stage to random will never reveal results as accurate as when using a fixed stage progress.

Table 7.2: Percentage deviation from the real PMI calculated by the new model (Ml) and the ADH method (ADH) using the temperature data from the datalogger (l), the weather station (w) and the calculated values using a regression analysis (r), temp = mean temperature.

real PMI	run	Ml l	Ml w	Ml r	ADH l	ADH w	ADH r	temp
509 hrs	Pig 1	1.6%	7.5%	3.7%	-2.2%	4.7%	-2.6%	21.5°C
130 hrs	Pig 2	4.5%	17.2%	6.5%				21.9°C
408 hrs	Pig 3	16.6%	22.3%	14.7%	7.6%	8.3%	11%	9.6°C

The third piglet was exposed in October 2009, the mean temperature was 9.6 °C. After 408 hours, larvae were collected at the end of the third instar stage. Using temperature data recorded by the data logger, the new model calculated the age of a third instar larvae with the stage progress fixed at the end of the stage as 475.8 hours (SD = 15 hours). This is a percentage deviation of 16.6 % (Table 7.2). The result when using data from the weather station was 499.8 hours (SD = 28.9, deviation: 22.3 %) and after using the regression based temperature data it was 469.9 (SD = 22.5, deviation: 14.7 %). The results of the ADH method were as follows: actual temperature data: 439 hours (deviation: 7.6 %), data from the weather station: 442 (deviation: 8.3 %) and data calculated after applying a regression formula: 453 hours (deviation: 11 %). The regression formula for this run was as follows:

$$\text{Temperature} = -2.34 + (1.21 * x) \tag{7.3}$$

The estimations based on the ADH method were better than those calculated by the new model, although the temperature was in a range where the ADH method is supposed to be unreliable.

Fluctuating temperatures in the climate chamber

Additionally to constant temperatures, two runs were carried out under the influence of fluctuating temperatures in the climate chamber. The larvae experienced twelve hours of 15 °C and twelve hours of 25 °C. Under constant 20 °C completion of the life cycle of *C. vicina* took 564 hours (SD = 7.3). Under fluctuating temperatures development took 576 hours (SD = 2.8). The model calculated 558.6 hours (SD = 6.9). Using the ADH method, 540 hours were calculated.

Analysis of the percentage deviation (Table 7.3) reveals that the largest aberration

Table 7.3: Percentage deviation from the real PMI estimated under fluctuating temperatures (15 and 25 °C) of the values estimated at constant temperatures, calculated by the new model for and the ADH method for fluctuating temperatures (15 and 25 °C).

real PMI	constant temp	model calculation	ADH calculation
564 hrs	2.1%	-1%	-4.3%

was achieved when using the ADH method (-4.3 %). This is a difference of 24 hours. The result of the model's calculation is only 1 % deviation and therefore very close to the real PMI, closer then the time interval measured for constant 20 °C.

Real case

On the evening of July 1st, 2009 a mother reported her son missing. She has seen him in the morning of July 1st, 2009 for the last time. In his room she found a suicide note. On the afternoon of July, 3rd 2009 the boy was found dead in a nearby forrest. The death was due to hanging. From the time he has last been seen till the time of finding 55 hours had past. The corpse was transported to the Institute of Forensic Medicine instantly and kept in a cooling chamber (4 °C) till July 7th, 2009 when he was autopsied (144 hours after he has last been seen alive). During autopsy, larvae of *C. vicina* were collected, killed with boiling water and preserved in 80 % EtOH. The most-developed larvae measured 12.5mm (3rd instar larvae). The temperature values were taken from the nearest weather station (Institute of Meteorology, University of Bonn), recorded in hourly intervals. An inspection of the finding place was not conducted, so that no reference temperature data could be collected. The temperature data from the weather station for the missing interval and from the cooling device were inserted in the model. The temperature error was set to 1 °C. The progress of the developmental stage was set to 0.1 as a length of 12.5 mm only occurs at the beginning of the 3rd instar stage (Fig. 7.5). The model calculated a PMI of 151.5 hours (SD = 7.5 hours). This is slightly longer than the missing interval (144 hours) but the correct result lies within the standard deviation. For the calculation of the PMI with an alternative method only the developmental data from Reiter can be consulted (Reiter, 1984). There were two possibilities for the calculation of the PMI using Reiters development table: either one assumed that below 4 °C no development took place and the mean temperature value for the time the corpse was hanging in the forrest (23 °C) was used. Reiter noted 66 hours at 22-23 °C till the larvae measured 12 mm. This would indicate a oviposition time when the person was still alive. The second possibility was to calculate a mean temperature value for the hanging time plus the cooling period: 12 °C. According to Reiter, development till a length of 12 mm under constant 12 °C takes 216 hours. This time interval is also too long. To sum up, when using the data of Reiter it was not possible to calculate a PMI that fitted the missing interval, the calculations always lead to a oviposition time where the boy was still alive. The model also calculated a slightly longer interval, which is nevertheless correct within the given standard deviation.

7.4 Discussion

Calculation of the post-mortem interval (PMI) is one of the most important applications in forensic entomology. The age of the most developed blow fly larvae can give reasonable information about the minimum time a person has been dead (Catts and Goff, 1992). Two different methods of larval age calculation are frequently used: isomegalen diagrams and ADH method. In both methods it is not possible to estimate standard deviations or the goodness of the calculation. Furthermore, the ADH method is only reliable between certain temperature thresholds where the developmental rate behaves linear. A data set of developmental values for *L. sericata* (Grassberger and Reiter, 2001) was used to develop a new model for the calculation of larval age (see previous chapter). When analyzing the developmental rate of *L. sericata* based on the data of Grassberger and Reiter the linear part lies between 18.5 and 22 °C (Fig. 7.1 B). Obviously, the temperature range in Central Europe is much wider. The new model is based on the actual developmental curve for each of the developmental stages by applying an exponential function (Fig. 7.2 A). The form of the exponential function is the base of the program but data sets for the development of different species can be put in. For each data set the exponential function is fitted newly and the calculation is based on the values of the selected data set (Fig. 7.2 B, 7.6). The model is designed to include sources of error such as temperature inaccuracies and to calculate a standard deviation. To validate the model's goodness developmental data for blow flies (*L. sericata* and *C. vicina*) originating from Bonn were produced and implemented in the model. The data of *L. sericata* were processed in the Diploma thesis of Sarah Gulinski. The data of *C. vicina* were generated and validated during this work. Generation of the developmental data occured under five constant temperatures: 13, 19, 20, 21 and 25 °C. Validation of the data was performed by exposing three dead piglets with egg batches from *C. vicina* under natural temperature regimes. In different stages of the development specimens were preserved as if they were material from a real case and the PMI was calculated using the new model and also the ADH method. In one case, development was completed, in one case the progress of the developmental stage (3rd instar) was unknown and in the third case the larvae developed to the end of the third larval stage. In the first two runs the mean temperature was about 21 °C and in the third run it was below 10 °C. Although only 5 data points were experimentally determined, the results of the model's calculations were good. In the first run, development till eclosion was completed, the time of hatching of the adult flies in the experimental box was observed with an accuracy of a few hours. A comparison of the calculation of the developmental time by the new model and the ADH method reveals that the new model gave the better result. The deviation from the real developmental time was 1.6 % for the model and -2.2 % for the ADH method. Nevertheless, results that good could only be achieved when the actual temperatures recorded by the dataloggers were used. Using temperature values from the nearest weather station for both methods resulted in deviations between 5 and 7 %. However, also good results were achieved when determining a regression formula based on temperature data recorded on the actual exposure site for 24 hours and the corresponding data from the weather station and applying the formula to the data from the weather station for the complete exposure interval. When using the regression based temperatures, the ADH method achieved better results than the model (-2.6 % deviation for the ADH method, 3.7 % for the new model) but the calculated PMI was too short. In this case the piglet was exposed 21 days till adult flies were observed (morning till morning). The model's calculation resulted in 21.5 days, so that in an official report the correct day would still be stated. The ADH method however, resulted in 20.5 days, which lead to a probable egg deposition time of 11:30 p.m. As blow flies will not oviposit during night-time (Amendt et al., 2008) one would probably conclude that oviposition took place in the morning of the next day. This decision would obviously increase the percentage deviation from the real PMI. In general, the use of five data points as input in the model seems to give reliable statements about the PMI, although the developmental rate (Fig. 7.4 B) did not result in a curve shape proposed for poikilothermically development. This is due to the fact that the model uses the form of the exponential function as developed using the data of *L. sericata* (Grassberger and Reiter, 2001) and fits the function to the implemented data, in this case Table 7.1 for *C. vicina*. The ADH method also gave reasonable results. This was expected as the experimental temperature range matched the temperature span in which the ADH method can be applied.

The second piglet was exposed under similar temperatures as the first one (mean temperature: 21.9 °C). However, the larvae of question were collected during the 3rd instar stage. This means that the progress of the stage was unknown. For the program it is possible to include the progress of the stage as random but for the ADH method it is not. One way would be to breed the larvae at a known temperature till the end of the stage or the end of development. If that is not possible because the larvae were already preserved one can calculate the age by using the literature value for the end of the stage and consider the introduced uncertainty as such. Another way is to check the literature for other developmental data. Both methods are not very accurate, as to be seen in the results. In the example of the Reiter data, one uncertainty is introduced as he gives a range between 22 and 23 °C as temperature value. As discussed previously, a difference of 1 °C has a huge impact on the developmental duration. As the collected larvae did not finish the 3rd instar when preserved, PMI estimation using the ADH method was neglected. The model however, can include the progress of the larval stage as random, which is a big advantage. The resulting standard deviation is consequently increased but a reasonable statement can still be given. In the example of pig 2, the calculated larval age deviated in a 5 % range from the actual PMI. However, using temperature data from the weather station increases the deviation to 17 %. So also in this case the accuracy is higher when using a regression analysis and adjusting the temperature values from the weather station to the actual finding place.

The model was originally designed to assure higher accuracy in temperature regimes the ADH method is not reliable. The third piglet was exposed under a mean temperature regime of 9.6 °C. The results of the calculations showed however that the percentage deviations from the real PMI of the estimations based on the new model were higher than of the results of the ADH method. Nevertheless, the deviations of the results calculated with the ADH method were higher than for the other temperatures. The lowest temperature that was applied to generate the developmental data of *C. vicina* was 13 °C. The originally implemented data of *L. sericata* did not include temperatures lower than 15 °C. As a result, the model has to extrapolate the values for colder temperatures. The big advantage of the model is that data for lower temperatures can be generated and additionally inserted to improve the results given by the model. For the ADH method this is not possible. Nevertheless, also in this trial an improvement can be noted when using the regression formula to adjust the data from the weather station, although the best results are always achieved when using the actual temperature data recorded at the experimental scene.

The analysis of the data generated under fluctuating temperatures emphasizes the problem discussed in chapter 1.4.2: the complete developmental time differs from the developmental time estimated at the corresponding mean temperature value. A comparison of the values of the percentage deviation from the actual measured time interval from egg till eclosion shows that the calculation of the PMI using the new model gives the best results, it differs only by 1 % although the data points were estimated under constant temperatures. The result of the ADH calculation differs by nearly 5 %. This was expected as the temperature reached values below the area in which it is safe to use the ADH method.

Evaluation of a real case in which the missing interval was probably identical to the post-mortem interval revealed that the use of the new model lead to much more accurate results than using other developmental data. The percentage deviation between the result calculated by the new model and the real PMI was 5 %. The calculated PMI values based on the Reiter data deviated up to 20 % and 50 %, respectively, from the real PMI. These are severe differences which indicate strongly the use of the new model. To estimate the progress of the developmental stage the length of the larvae was used and related to the scatter plot showing all sizes occurring during the 3rd instar stage. In this case it was safe to conclude that a 12 mm long larvae had entered the 3rd stage recently. In the case of piglet 2 a 18 mm long larvae was collected. Figure 7.5 shows the wide span of 3rd instar larvae measuring 18 mm, therefore the calculation was conducted with a random progress of the stage.

When calculating the PMI in a real case, no regression based temperature data could be used since no inspection of the finding place was carried out. Nevertheless, the calculation fitted well, in contrast to the piglet trials, where the adjusted data were the better choice in all cases. This might be because the corpse was exposed in a shaded, sun protected area whereas the piglets were exposed on the roof of the institute. The sun will shine on the roof for most part of the day and generate more heat than at the place where the weather station records the temperature. The meteorological station of the University of Bonn states on their webpage that according to the choice of the location a systematical error is introduced in the measurement of the temperature as a shading due to adjacent buildings and vegetation occurs in the afternoon. These microclimatically differences between the weather station and the finding place of a corpse can have severe effects as to be seen in the piglet trials. The best way is to visit the finding place and to take temperature measurements on site. Then it can be assured that the microclimatically differences can be recognized and considered in the following calculation.

Conclusion:

- The main reason for creating a new model was to establish a method that gives a standard deviation for the PMI calculation. The SDs as calculated for the experimental trials provided a reasonable range the real PMI lied within. As discussed, the SD increases when the progress of the stage is unknown. Therefore it is important to establish a method that determines the progress dependable or one should breed the collected maggots till eclosion under known temperatures.

- Temperature data from a weather station nearby the collection scene must be considered carefully. Instead of using the data plainly, one could record the temperature for 24 hours at the collection scene and apply a regression analysis to adjust the data from the weather station. If this is not possible the temperature error for the model's calculation should be increased. Furthermore, in cases where larvae are collected indoors it is important to know how the strongly temperature fluctuated. As to be seen in the PMI calculation in a real case, the microclimatical differences between weather station and finding situation must be considered and if they do not differ dramatically, using these data will lead to good results.

- Even in temperature regimes where the ADH method is supposed to be reliable, the new model achieved slightly better results.

- In temperature regimes beyond the reliable ADH range the model does not yet achieve better results but can be improved by inserting more data points extended to colder temperatures.

- Fluctuating temperatures are considered much more reliable by the new model than by the ADH method. The new model does not use the corresponding mean value but applies for every developmental step the real temperature. This leads to much better results.

- A big advantage is that every data set can be implemented in the model. Every forensic entomologist can use his or her own developmental data estimated for their geographical region.

- Nevertheless, a lot of additional data must be produced to improve the model in colder temperature regions and more experiments for the validation must be conducted.

Part V

Discussion

Chapter 8

Discussion

Forensic entomology is a research field that might be considered both old and new. The observation of insects infesting decomposing material has probably been made since the existence of mankind. The application of this observation in investigation of crime cases, however, is still a new discipline and has huge research potential. Knowledge about behavior, development and colonization patterns of corpse associated insects form the basis for forensic entomology.

8.1 Ecology: how environmental factors dictate behavior

Ecology can be considered as a description of interactions between organisms and their environment (Speight et al., 2008). In forensic entomology, arthropods found in close proximity to a corpse or a crime scene are investigated in relation to the environmental factors that influenced their behavior. One of these environmental factors is the location: in a wider sense the country or the region and at a smaller scale micro-climatic regions such as indoors or outdoors, bright sunlight or a shaded area. Other factors are the climatical circumstances, interactions with other organisms and the decomposition stage of the primary habitat: the corpse. All these influences have to be factored in when conclusions are to be drawn about e.g. a post-mortem interval based on entomological evidence.

The location in a wider sense (country or region) is important as some insect species do not occur world-wide. Some species that are very commonly found on carcass in the United States as *L. coeruleiviridis* (Tabor et al., 2004) or *Chrysomya*

8. Discussion

rufifacies (Shahid et al., 2000) are not present in Central Europe. Thus, correct species identification followed by a survey about the distribution of the identified species are crucial in working with real cases. The location in a narrower sense is important as the colonization time by arthropods can vary. As shown in chapter 4, carcasses exposed indoors are infested by blow flies with a delay of up to 24 hours. Obviously, for the trials only one type of house in one area was used and also only small carcasses (dead piglets). Nevertheless, Anderson showed recently that a larger pig carcass (42 kg) was infested up to 5 days later when being exposed indoors compared to outdoors (Anderson, 2009). The experiments were conducted in Canada in late May and the piglets were clothed. The experimental setup however did not include a tilted window, instead two windows were open (that is the lower part is pushed upwards underneath the upper part) but screened and one screen had a small hole in the mesh. Although the piglets were larger than the ones used in the experiments for this thesis they were not infested faster in the indoor location. It seems as if the access for the flies is the limiting factor and not the arising odors since larger carcasses should produce more smell. Also, open but screened windows will allow the smell to diffuse easily. Nevertheless, access through a small hole in the screening seems for blow flies to be more difficult than entering through a tilted window as the flies in Anderson's trials arrived several days later than in the experiments for this thesis. Moreover, it seems that a lot of factors influence the infestation time in indoor scenarios and although it is safe to say that indoor carcasses will be infested with a delay it is necessary to conduct much more experiments to test for more variables. Moreover, not only the behavior of adult blow flies is influenced by the location but also the behavior of the larvae. Outdoors, wandering larvae can proceed for several meters till they find a suitable place for pupation. Indoors, they are limited by walls, carpets and furniture and might pupate faster than outdoors. Certainly, pupal cases are more easily found indoors than outdoors. Finishing the thoughts on the impact of the location, it seems also of importance how cultural scenarios influence experimental designs. As shown in the comparison of the indoor trials in Canada and Germany, it becomes obvious that the concept of a window for example is not the same in every part of the world and can therefore lead to different results in similar experiments.

But not only an indoor scenario can delay the infestation with blow fly eggs; also wrapping the corpse or storage in a trash can be an obstacle for arthropods. For a

PMI calculation it is nonetheless inevitable to determine the time interval the insects took to reach the corpse as accurately as possible. Not many experiments have been conducted to cover that problem. The indoor/outdoor experiments designed for this thesis were among the first and their results show their importance: an unrecognized delay in insect infestation can lead to severe miscalculations of the PMI. Moreover, the list of finding situations of corpses can be continued easily: buried corpses (different depths) (Amorim and Ribeiro, 2001; Merritt et al., 2007; Turner and Wiltshire, 1999; VanLaerhoven and Anderson, 1999), bodies in water (Haskell et al., 1989; Hobischak and Anderson, 2002; Tomberlin and Adler, 1998), bodies in vehicles (Voss et al., 2008), burned bodies (Avila and Goff, 1998; Pai et al., 2007) and bodies in garbage bins (Reibe et al., 2008). As shown in the case of the latter scenario the main question that arose concerned the accessibility of the corpse for insects and the consequences for the estimation of the PMI.

Temperature as an environmental factor influences the developmental behavior of blow fly larvae dramatically. Therefore, the microclimate of each location is of great importance and must be determined as accurately as possible as it influences temperature values the developing larvae experience (see 8.2). Unfortunately, it is sometimes difficult to reconstruct the climate at the finding place of a corpse and one has to work with assumptions and approximations, for example when recording temperature values after the finding and relating those to values recorded at the nearest weather station. In the same manner as the larval development is influenced by temperature, the decomposition rate is temperature dependent too. As shown in chapter 1, different decomposition stages attract different insect species. Therefore, to reconstruct a timeline since death, it is most important to include all environmental factors dictating arthropod behavior and development.

8.2 Development: calculating larval age with the devil in the detail

The application used most in forensic entomology is the estimation of the PMI based on the developmental progress of blow fly larvae developing on the corpse in question. As shown in chapter 1, several methods are used to calculate the larval age based on the progress of the development and the temperature that influenced the larvae during their development. In chapter 6, a new model to calculate larval age

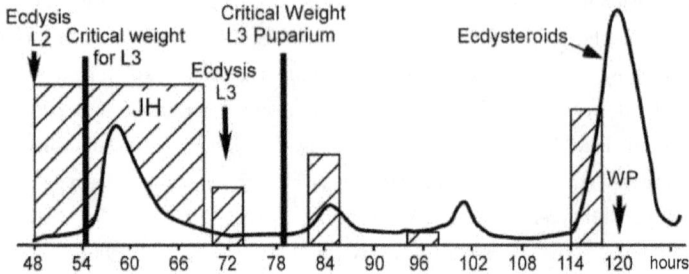

Figure 8.1: Development of *D. melaogaster* and corresponding hormone titers between 2nd larval stage and onset of pupariation (white puparium = WP) (from Riddiford et al. 2003)

was introduced based on the actual non-linear development curves. The method was validated using developmental data for *L. sericata* and *C. vicina* originating from Bonn, Germany and gave rather good results (chapter 7). All calculations of a PMI were reasonable. In the recent published 2nd edition of the book "Forensic Entomology. The Utility of Arthropods in Legal Investigations" in the chapter on insect development it is said that curvilinear models have no improvement over linear models. It is also said that perhaps the failure of curvilinear models is related to accumulation of errors in the estimates of parameters needed to drive such models (Higley and Haskell, 2009). However, as shown in chapter 7, the calculations of a PMI by the new curvilinear model achieved in most cases more accurate results than the calculations by the non-linear ADH method.

One difficulty in using the model properly, apart from the correct determination of the temperature values that actually influenced the developmental rate, is the determination of the progress within the developmental stage. One way certainly is to observe the further development of collected larvae under known temperatures. If that is not possible, the model allows to include the progress as random which leads to results with an enlarged standard deviation. However, it might be useful to find a method that allows correct determination of the stage progress of the collected larvae. As shown in chapter 1, development is under control of several hormones. In Figure 8.1 the hormone titers of juvenile hormone (JH) and ecdysteroids during the development of *D. melanogaster* are shown in relation to the progress of the 2nd and 3rd developmental stage (Riddiford et al., 2003). It might be a reasonable

114

idea to estimate the concentration of mRNA of certain developmental genes during larval development to refine the determination of the progress of each developmental stage.

8.3   Cases

The choice of the Institute of Forensic Medicine as research facility brought the benefit of real case work. About 65 cases involving insect material were collected and evaluated. In 18 cases the corpse was found outdoors. Indoor cases were often very much alike: A socially isolated person died alone in his/her apartment and was found several days or weeks later in an advanced stage of decay as well as blow fly larvae infestation. For all indoor cases it was impossible to verify the calculated PMI as nobody reported the persons missing. As shown in chapter 4, the best results when working on PMIs of indoor corpses could be retrieved when specimens of Phoridae infested the corpse as they often reached it faster than the blow flies (Reibe and Madea, 2010b).

The blow fly species found in all cases were *L. sericata*, *C. vicina*, *P. terraenovae* and rarely *C. vomitoria* and *L. caesar*. *L. illustris* which was caught during the indoor/outdor experiments was never observed in a real case.

In all outdoor cases the insect evidence proved to be a good complement to other methods used by the police and medico-legal doctors. In one case, an accidental death of a mentally sick person, the police extended their search in the missing person's reports to one year earlier than indicated by the medico-legal doctors due to my report on the insect evidence and could identify the person eventually. In other cases, the insect evidence could confirm results of the police investigations (Reibe et al., 2008).

The outdoor cases in contrast to the indoor cases were all very different. In two of the outdoor cases someone was murdered, in several cases it was suicide or an accident. However, the circumstances, the finding situations, the climate, the season and the missing intervals were all very different. That made it difficult on the one hand to evaluate all factors, as discussed previously, but on the other hand it accounted for the magic of the field: using biological evidence to solve a riddle.

8.4   Conclusion and Outlook

Forensic entomology is a promising field of research that will hopefully become implemented in routine case work in institutes of forensic medicine. Although several aspects still have to be researched, the basics are useful tools in post-mortem interval estimation. Knowledge from physiology and ecology of insects (Reibe and Madea, 2010a) can be coupled with genetic tools (Reibe et al., 2009) to improve both basic research and application in real case work.

Bibliography

Adams Z and Hall M (2003) Methods used for the killing and preservation of blowfly larvae, and their effect on post-mortem larval length. Forensic Sci Int 138(1-3):50–61

Altschul SF, Gish W, Miller W, Myers EW and Lipman DJ (1990) Basic local alignment search tool. J Mol Biol 215(3):403–410

Amendt J, Krettek R, Niess C, Zehner R and Bratzke H (2000) Forensic entomology in Germany. Forensic Sci Int 113(1-3):309–314

Amendt J, Krettek R and Zehner R (2004) Forensic entomology. Naturwissenschaften 91(2):51–65

Amendt J, Zehner R and Reckel F (2008) The nocturnal oviposition behaviour of blowflies (Diptera: Calliphoridae) in Central Europe and its forensic implications. Forensic Sci Int 175(1):61–64

Ames C and Turner B (2003) Low temperature episodes in development of blowflies: implications for postmortem interval estimation. Med Vet Entomol 17(2):178–186

Ames C, Turner B and Daniel B (2006) The use of mitochondrial cytochrome oxidase I gene (COI) to differentiate two UK blowfly species -Calliphora vicina and Calliphora vomitoria. Forensic Sci Int 164:179–182

Amorim JA and Ribeiro OB (2001) Distinction among the puparia of three blowfly species (Diptera: Calliphoridae) frequently found on unburied corpses. Mem Inst Oswaldo Cruz 96(6):781–784

Anderson G (2001) *Forensic Entomology: The Utility of Arthropods in Legal Investigations*, chapter Insect Succession on Carrion and its Relationship to Determining Time of Death. CRC Press, pp. 143–175

BIBLIOGRAPHY

Anderson G (2009) *Forensic Entomology: The Utility of Arthropods in Legal Investigations 2nd ed.*, chapter Factors that influenece insect succession on carrion. CRC Press, pp. 201–250

Anderson GS and VanLaerhoven SL (1996) Initial Studies on Insect Succession on Carrion in southwestern British Columbia. Jounal of Forensic Science 41:617–625

Archer M (2003) Annual variation in arrival and departure times of carrion insects at carcasses: Implications for Succession Studies in Forensic Entomology. Australian Journal of Zoology 51:569–576

Archer MS (2004a) The effect of time after body discovery on the accuracy of retrospective weather station ambient temperature corrections in forensic entomology. J Forensic Sci 49(3):553–559

Archer MS (2004b) Rainfall and temperature effects on the decomposition rate of exposed neonatal remains. Sci Justice 44(1):35–41

Archer MS, Bassed RB, Briggs CA and Lynch MJ (2005) Social isolation and delayed discovery of bodies in houses: the value of forensic pathology, anthropology, odontology and entomology in the medico-legal investigation. Forensic Sci Int 151(2-3):259–265

Archer MS and Elgar MA (2003) Yearly activity patterns in southern Victoria (Australia) of seasonally active carrion insects. Forensic Sci Int 132(3):173–176

Arnaldos I, Romera E, García MD and Luna A (2001) An initial study on the succession of sarcosaprophagous Diptera (Insecta) on carrion in the southeastern Iberian peninsula. Int J Legal Med 114(3):156–162

Arnott S and Turner B (2008) Post-feeding larval behaviour in the blowfly, Calliphora vicina: effects on post-mortem interval estimates. Forensic Sci Int 177(2-3):162–167

Avila FW and Goff ML (1998) Arthropod succession patterns onto burnt carrion in two contrasting habitats in the Hawaiian Islands. J Forensic Sci 43(3):581–586

Beattie MVF (1928) Observations on the thermal death points of the blow-fly at different relative humidities. Bull Ent Res 18:397–403

Beck SD (1950) Nutrition of the European corn borer, Pyrausta nubilalis (Hbn.). II. Some effects of diet on larval growth characteristics. Physiol Zool 23(4):353–361

Benecke M (1998) Random amplified polymorphic DNA (RAPD) typing of necrophageous insects (Diptera, Coleoptera) in criminal forensic studies: validation and use in practice. Forensic Sci Int 98(3):157–168

Benecke M (2001) A brief history of forensic entomology. Forensic Sci Int 120(1-2):2–14

Benecke M (2005) *Forensic Pathology Reviews, Vol. 2*, chapter 10 Arthropods and Corpses. Humana Press Inc., pp. 207–240

Benecke M (2008) A brief survey of the history of forensic entomology. Acta biologica Benrodis 14:15–38

Benecke M, Josephi E and Zweihoff R (2004) Neglect of the elderly: forensic entomology cases and considerations. Forensic Sci Int 146 Suppl:S195–S199

Benecke M and Lessig R (2001) Child neglect and forensic entomology. Forensic Sci Int 120(1-2):155–159

Benecke M and Seifert B (1999) [Forensic entomology exemplified by a homicide. A combined stain and postmortem time analysis]. Arch Kriminol 204(1-2):52–60

Bharti M and Singh D (2003) Insect faunal succession on decaying rabbit carcasses in Punjab, India. J Forensic Sci 48(5):1133–1143

Boehme P, Ament J and Zehner R (2009) Using real time PCR approaches to determine a blowfly pupas age. In *Abstractbook of the EAFE 2009*

Bowler K and Terblanche JS (2008) Insect thermal tolerance: What is the role of Ontogeny, Ageing and Senescence? Biol Rev Camb Philos Soc

Browne LB, Bartell RJ and Shorey HH (1969) Pheromone-mediated behaviour leading to group oviposition in the blowly Lucilia cuprina. Jounal Insect Physiol 15:1003–1014

Büchner F (1965) *Pläne und Fügungen. Lebenserinnerungen eines deutschen Hochschullehrers.* Urban und Schwarzenberg

Campobasso, Disney and Introna (2004) A case of Megaselia scalaris (Loew) (Dipt., Phoridae) breeding in a human corpse. Aggrawal's Internet Journal Forensic of Forensic Medicine and Toxicology 5:3–5

Catts EP and Goff ML (1992) Forensic entomology in criminal investigations. Annu Rev Entomol 37:253–272

Charabidze D, Bourel B, Hedouin V and Gosset D (2009) Repellent effect of some household products on fly attraction to cadavers. Forensic Sci Int 189(1-3):28–33

Chen WY, Hung TH and Shiao SF (2004) Molecular identification of forensically important blow fly species (Diptera: Calliphoridae) in Taiwan. J Med Entomol 41(1):47–57

Davies L and Ratcliffe GG (1994) Development rates of some pre-adult stages in blowflies with reference to low temperatures. Med Vet Entomol 8(3):245–254

Dekeirsschieter J, Verheggen FJ, Gohy M, Hubrecht F, Bourguignon L, Lognay G and Haubruge E (2009) Cadaveric volatile organic compounds released by decaying pig carcasses (Sus domesticus L.) in different biotopes. Forensic Sci Int 189(1-3):46–53

Denlinger DL (1994) Metamorphosis behavior of flies. Annu Rev Entomol 39:243–266

Desmyter S and Gosselin M (2009) COI sequence variability between Chrysomyinae of forensic interest. Forensic Sci Int Genet 3(2):89–95

Disney R (1989) *Scuttle Flies: The Phoridae (Genius Megaselia). The Royal Entomological Society Handbooks, Vol. 10, Part 8.* The Royal Entomological Society, London

Disney R (1994) *Scuttle flies: the Phoridae.* Chapman and Hall, London/UK

Disney RHL (2005) Duration of development of two species of carrion-breeding scuttle flies and forensic implications. Medical and Veterinary Entomology 19:229–235

Disney RHL (2008) Natural history of the scuttle fly, Megaselia scalaris. Annu Rev Entomol 53:39–60

BIBLIOGRAPHY

Disney RHL and Manlove JD (2005) First occurrences of the Phorid, Megaselia abdita, in forensic cases in Britain. Med Vet Entomol 19(4):489–491

Donovan SE, Hall MJR, Turner BD and Moncrieff CB (2006) Larval growth rates of the blowfly, Calliphora vicina, over a range of temperatures. Med Vet Entomol 20(1):106–114

Eberhardt TL and Elliot DA (2008) A preliminary investigation of insect colonisation and succession on remains in New Zealand. Forensic Sci Int 176(2-3):217–223

Erzinclioglu YZ (1990) On the interpretation of maggot evidence in forensic cases. Med Sci Law 30(1):65–66

Erzinclioglu Z (1996) Blowflies. The Richmond Publishing Co. Ltd

Fabritius K and Klunker R (1991) Die Larven- und Puparienparasitoide von synanthropen Fliegen in Europa. 32(1):1–24

Faucherre J, Cherix D and Wyss C (1999) Behavior of Calliphora vicina (Diptera, Calliphoridae) under extreme conditions. Journal of Insect Behavior 12(5):687–690

Feist (1926) Verhalten von Fliegenmaden gegenüber Giften. Zeitschr f Untersuchung der Lebensmittel :466–469

Field A (2005) Discovering Statistics Using SPSS. SAGE Publications

Goff M (1993) Estimation of Postmortem Interval Using Arthropod Development and Successional Patterns. Forensic Science Review 5 (2):81–94

Goff M, Odom C and Early M (1986) Estimation of Post-Mortem interval by entomological techniques: A case study from oahu, Hawaii. BullSocVector Ecol :242–246

Goff ML (1991a) Comparison of Insect Species Associated with Decomposing Remains Recovered Inside Dwellings and Outdoors on the Island of Oahu, Hawaii. Journal of Forensic Science 36(3):748–753

Goff ML (1991b) Comparison of insect species associated with decomposing remains recovered inside dwellings and outdoors on the island of Oahu, Hawaii. J Forensic Sci 36(3):748–753

Gomes L, Gomes G, Oliviera H, Sanches M and Zuben CV (2006) Influence of photoperiod on body weight and depth of burrowing in larvae of *Chrysomya megacephala* (Fabricius) (Diptera, Calliphoridae) and implications for forensic entomology. Revista Brasileira de Entomologia 50(1):76–79

Grassberger M and Frank C (2004) Initial study of arthropod succession on pig carrion in a central European urban habitat. J Med Entomol 41(3):511–523

Grassberger M and Reiter C (2001) Effect of temperature on Lucilia sericata (Diptera: Calliphoridae) development with special reference to the isomegalen- and isomorphen-diagram. Forensic Sci Int 120(1-2):32–36

Grassberger M and Reiter C (2002) Effect of temperature on development of the forensically important holarctic blow fly Protophormia terraenovae (Robineau-Desvoidy) (Diptera: Calliphoridae). Forensic Sci Int 128(3):177–182

Greenberg B (1990) Nocturnal Oviposition Behavior of Blow Flies (Diptera: Calliphoridae). J Med Entomol 27 (5):807–810

Greenberg B (1991) Flies as forensic indicators. J Med Entomol 28(5):565–577

Greenberg B and Wells JD (1998) Forensic use of Megaselia abdita and M. scalaris (Phoridae: Diptera): case studies, development rates, and egg structure. J Med Entomol 35(3):205–209

Greene G (1996) Rearing Techniques for Creophilus Maxillosus, a Predator of Fly Larvae in Cattle Feedlots. Journal of Economic Entomology 89(4):848–851

Hajibabaei M, Singer GAC, Hebert PDN and Hickey DA (2007) DNA barcoding: how it complements taxonomy, molecular phylogenetics and population genetics. Trends Genet 23(4):167–172

Harvey ML, Gaudieri S, Villet MH and Dadour IR (2008) A global study of forensically significant calliphorids: implications for identification. Forensic Sci Int 177(1):66–76

Harvey ML, Mansell MW, Villet MH and Dadour IR (2003) Molecular identification of some forensically important blowflies of southern Africa and Australia. Med Vet Entomol 17(4):363–369

Haskell N and Catts E, editors (1990) *Entomology and Death: a Procedual Guide.* Forensic Entomology Associates

Haskell NH, McShaffrey DG, Hawley DA, Williams RE and Pless JE (1989) Use of aquatic insects in determining submersion interval. J Forensic Sci 34(3):622–632

Hewadikaram KA and Goff ML (1991) Effect of carcass size on rate of decomposition and arthropod succession patterns. Am J Forensic Med Pathol 12(3):235–240

Higley L and Haskell N (2009) *Forensic Entomology, The Utility of Arthropods in Legal Investigations 2nd ed.*, chapter Insect Development and Forensic Entomology. CRC Press LLC, pp. 389–405

Hobischak NR and Anderson GS (2002) Time of submergence using aquatic invertebrate succession and decompositional changes. J Forensic Sci 47(1):142–151

Hobson R (1932) Studies on the nutrition of blow-fly larvae. J Exptl Biol 9:359–365

Hoffmann K (1995) *Physiologie der Insekten.* Gustav Fischer

Honek A (1996) Geographical variation in thermal requirements for insect development. European Journal of Entomology 93:303–312

Horenstein M, Linhares A, Rosso B and Garcia M (2007) Species composition and seasonal succession of saprophagous calliphorids in a rural area of Córdoba, Argentina. Biol Res 40:163–171

Janisch E (1928) Die Leben- und Entwicklungsdauer der Insekten als Temperaturfunktion. Zeitschrift für wissenschaftliche Zoologie 132:176–186

Janisch E (1931) Experimentelle Untersuchungen über die Wirkung der Umweltfaktoren auf Insekten. Z Morph u Ökol Tiere 22:287–348

Joy JE, Liette NL and Harrah HL (2006) Carrion fly (Diptera: Calliphoridae) larval colonization of sunlit and shaded pig carcasses in West Virginia, USA. Forensic Sci Int 164(2-3):183–192

Kalinová B, Podskalská H, Rzicka J and Hoskovec M (2009) Irresistible bouquet of death–how are burying beetles (Coleoptera: Silphidae: Nicrophorus) attracted by carcasses. Naturwissenschaften 96(8):889–899

Kaneshrajah G and Turner B (2004) Calliphora vicina larvae grow at different rates on different body tissues. Int J Legal Med 118(4):242–244

Kaufmann O (1932) Einige Bemerkungen über den Einfluss von Temperaturschwankungen auf die Entwicklungsdauer und Streuung bei Insekten und seine graphische Darstellung durch Kettenlinie und Hyperbel. Z Morph Okol Tiere 25:353–361

Keh B (1985) Scope and applications of forensic entomology. Annu Rev Entomol 30:137–154

Lane RP (1975) An investigation into blowfly (Diptera: Calliphoridae) succession on corpses. J Nat Hist 9:581–588

Leclercq M (1983) [Entomology and forensic medicine: dating of a death. Unpublished case]. Rev Med Liege 38(19):735–738

Leclercq M (1988a) [Entomology applied to legal medicine. Its origins, its evolution]. Acta Med Leg Soc (Liege) 38(1):225–232

Leclercq M (1988b) [How to carry out an entomologic medico-legal assessment]. Acta Med Leg Soc (Liege) 38(1):241–245

Leclercq M (1988c) [Medico-legal entomologic assessment: several examples]. Acta Med Leg Soc (Liege) 38(1):247–255

Leclercq M and Brahy G (1990) [Entomology and legal medicine: origins, development, actualization]. Rev Med Liege 45(7):348–358

Leclercq M, Dodinval P, Piette P and Verstraeten C (1991) [Example of team work between forensic medicine, odontology and entomology. Identification of human bones, dating of death and establishing the crime location]. Rev Med Liege 46(11):583–591

Lowne B (1890) The anatomy, physiology, morphology and development of the blowfly. RH Porter, London

Malgorn Y and Coquoz R (1999) DNA typing for identification of some species of Calliphoridae. An interest in forensic entomology. Forensic Sci Int 102(2-3):111–119

Manlove JD and Disney RHL (2008) The use of Megaselia abdita (Diptera: Phoridae) in forensic entomology. Forensic Sci Int 175(1):83–84

Mansson RA, Frey JG, Essex JW and Welsh AH (2005) Prediction of properties from simulations: a re-examination with modern statistical methods. J Chem Inf Model 45(6):1791–1803

Matthews R (1974) Biology of Braconidae. Annu Rev Entomol 19:15–32

Matuszewski S, Bajerlein D, Konwerski S and Szpila K (2008) An initial study of insect succession and carrion decomposition in various forest habitats of Central Europe. Forensic Sci Int 180(2-3):61–69

McDonagh L, Thornton C, Wallman JF and Stevens JR (2009) Development of an antigen-based rapid diagnostic test for the identification of blowfly (Calliphoridae) species of forensic significanc. Forensic Science Inernational: Genetics 3:162–165

Mellanby K (1938) Diapause and metamorphosis of the blowfly, Lucilia sericata Meig. Parasitology, Cambridge 30:392–402

Merritt RW, Snider R, de Jong JL, Benbow ME, Kimbirauskas RK and Kolar RE (2007) Collembola of the grave: a cold case history involving arthropods 28 years after death. J Forensic Sci 52(6):1359–1361

Moore H (2009) Usefulness of Hydrocarbons within Forensic Entomology in establishing the Postmortem Interval (PMI). In Abstractbook EAFE 2009

Nelson LA, Wallman JF and Dowton M (2007) Using COI barcodes to identify forensically and medically important blowflies. Med Vet Entomol 21(1):44–52

Nijhout HF (2008) Size matters (but so does time), and it's OK to be different. Dev Cell 15(4):491–492

Norris K (1965) The Bionomics of Blow Flies. Annual Review of Entomology 10:47–68

Nuorteva P (1965) The flying activity of blowflies (Dipt., Calliphoridae) in subarctic conditions. Ann Ent Fenn 31(4):242–245

Nuorteva P (1967) The synanthropy and bionomics of blowflies in subarctic Northern Finland. Wiad Parazytol 13(4):603–607

Nuorteva P, Schumann H, Isokoski M and Laiho K (1974) Studies on the possibilities of using blowflies (Dipt., Calliphoridae) as medicolegal indicators in Finland. Four cases where species identification was performed from larvae. Ann Entomol Fenn 40:70–74

Pai CY, Jien MC, Li LH, Cheng YY and Yang CH (2007) Application of forensic entomology to postmortem interval determination of a burned human corpse: a homicide case report from southern Taiwan. J Formos Med Assoc 106(9):792–798

Perotti MA and Braig HR (2009) Phoretic mites associated with animal and human decomposition. Exp Appl Acarol 49(1-2):85–124

Putmann R (1977) Dynamics of the Blowfly, *Calliphora erytrocephala*, within Carrion. J Anim Ecol 46:854–866

Redi F (1674) *Esperienze intorno alla generazione degl'insetti.* Francesco Onofri, Firenze

Reibe S and Madea B (2010a) How promptly do blowflies colonise fresh carcasses? A study comparing indoor with outdoor locations. Forensic Sci Int 195(1-3):52–57

Reibe S and Madea B (2010b) Use of Megaselia scalaris (Diptera: Phoridae) for post-mortem interval estimation indoors. Parasitol Res 106(3):637–640

Reibe S, Schmitz J and Madea B (2009) Molecular identification of forensically important blowfly species (Diptera: Calliphoridae) from Germany. Parasitol Res 106(1):257–261

Reibe S, Strehler M, Mayer F, Althaus L, Madea B and Benecke M (2008) [Dumping of corpses in compost bins–two forensic entomological case reports]. Arch Kriminol 222(5-6):195–201

Reiter C (1984) Zum Wachstumsverhalten der Maden der blauen Schmeißfliege *Calliphora vicina.* Z Rechtsmed 91:295–308

Richards CS, Crous KL and Villet MH (2009) Models of development for blowfly sister species Chrysomya chloropyga and Chrysomya putoria. Med Vet Entomol 23(1):56–61

BIBLIOGRAPHY

Riddiford LM, Hiruma K, Zhou X and Nelson CA (2003) Insights into the molecular basis of the hormonal control of molting and metamorphosis from Manduca sexta and Drosophila melanogaster. Insect Biochem Mol Biol 33(12):1327–1338

Rognes K (1991) *Blowflies (Diptera, Calliphoridae) of Fennoscandia and Denmark.* E. J. Brill/Scandinavian Science Press Ltd.

Roux O, Gers C and Legal L (2008) Ontogenetic study of three Calliphoridae of forensic importance through cuticular hydrocarbon analysis. Med Vet Entomol 22(4):309–317

Saigusa K, Takamiya M and Aoki Y (2005) Species identification of the forensically important flies in Iwate prefecture, Japan based on mitochondrial cytochrome oxidase gene subunit I (COI) sequences. Leg Med (Tokyo) 7(3):175–178

Saitou N and Nei M (1987) The neighbor-joining method: a new method for reconstructing phylogenetic trees. Mol Biol Evol 4 (4):406–425

Saunders D (1987) Maternal influence on the incidence and duration of larval diapause in *Calliphora vicina.* Physiol Entomol 12:331–338

Saunders D (1997) Under-seized larvae from short-day adults of the blow fly, *Calliphora vicina*, side-step the diapause programme. Physiol Entomol 22:249–255

Saunders D and Hayward S (1998) Geographical and diapause-related cold tolerance in the blow fly, Calliphora vicina. J Insect Physiol 44(7-8):541–551

Saunders DS, Macpherson JN and Cairncross KD (1986) Maternal and larval effects of photoperiod on the induction of larval diapause in two species of fly, Calliphora vicina and Lucilia sericata. Exp Biol 46(1):51–58

Schoenly KG, Haskell NH, Hall RD and Gbur JR (2007) Comparative performance and complementarity of four sampling methods and arthropod preference tests from human and porcine remains at the Forensic Anthropology Center in Knoxville, Tennessee. J Med Entomol 44(5):881–894

Schumann H (1965) Die Schmeißfliegengattung *Calliphora.* Merkblätter über angewandte Parasitenkunde und Schädlingsbekämpfung 6:1–14

Schumann H (1971) Die Gattung *Lucilia* (Goldfliegen). Merkblätter über angewandte Parasitenkunde und Schädlingsbekämpfung in Angewandte Parasitologie 18:1–20

Schumann H (1990) [The occurrence of Diptera in living quarters]. Angew Parasitol 31(3):31–41

Scott MP (1998) The ecology and behavior of burying beetles. Annu Rev Entomol 43:595–618

Shahid SA, Hall RD, Haskell NH and Merritt RW (2000) Chrysomya rufifacies (Macquart) (Diptera: Calliphoridae) established in the vicinity of Knoxville, Tennessee, USA. J Forensic Sci 45(4):896–897

Shahid SA, Schoenly K, Haskell NH, Hall RD and Zhang W (2003) Carcass enrichment does not alter decay rates or arthropod community structure: a test of the arthropod saturation hypothesis at the anthropology research facility in Knoxville, Tennessee. J Med Entomol 40(4):559–569

Sharanowski BJ, Walker EG and Anderson GS (2008) Insect succession and decomposition patterns on shaded and sunlit carrion in Saskatchewan in three different seasons. Forensic Sci Int 179(2-3):219–240

Sharpe P and DeMichele D (1977) Reaction Kinetics of Poikilotherm Development. J theor Biol 64:649–670

Sharpe P and Hu L (1980) Reaction kinetics of nutrition dependent poikilotherm development. J theor Biol 82:317–333

Shean BS, Messinger L and Papworth M (1993) Observations of differential decomposition on sun exposed v. shaded pig carrion in coastal Washington State. J Forensic Sci 38(4):938–949

Simpson K (1986) *Forty years of murder*. Grafton Books

Singh D and Bharti M (2001) Further observations on the nocturnal oviposition behaviour of blow flies (Diptera: Calliphoridae). Forensic Sci Int 120(1-2):124–126

Slone D and Gruner S (2007) Thermoregulation in larval aggregations of carrion-feeding blow flies (Diptera: Calliphoridae). J Med Entomol 44(3):516–523

Smirnov E and Zhelochovtsev A (1927) Veränderung der Merkmale bei Calliphora erythrocephala MG. unter dem Einfluß verkürzter Ernährungsperiode der Larve. Archiv für Entwicklungsmechanik 108:579–597

Smith K (1986) *A Manual of Forensic Entomology*. The Trustees of the Brisith Museum (Natural History), London

Smith KE and Wall R (1997) The use of carrion as breeding sites by the blowfly Lucilia sericata and other Calliphoridae. Med Vet Entomol 11(1):38–44

Speight M, Hunter M and Watt A (2008) *Ecology of Insects: Concepts and Applications*. John Wiley & Sons

Sperling FA, Anderson GS and Hickey DA (1994) A DNA-based approach to the identification of insect species used for postmortem interval estimation. J Forensic Sci 39(2):418–427

Steinborn H (1976) *Ökologische Untersuchungen an Schmeißfliegen (Calliphoridae)*. Master's thesis, Christian-Albrechts-Universität Kiel

Steinborn H (1981) Oekologische Untersuchungen an Schmeissfliegen (Calliphoridae). Drosera 81 (1):17–26

Steiner G (1948) Fallenversuche zur Kennzeichnung des Verhaltens von Schmeissfliegen gegenüber verschiedenen Merkmalen ihrer Umgebung. Zeitschrift für vergleichende Physiologie 31:1–37

Stevens J and Wall R (1996) Species, sub-species and hybrid populations of the blowflies Lucilia cuprina and Lucilia sericata (Diptera:Calliphoridae). Proc Biol Sci 263(1375):1335–1341

Stevens J and Wall R (1997) Genetic variation in populations of the blowflies Lucilia cuprina and Lucilia sericata (Diptera: Calliphoridae). Random amplified polymorphic DNA analysis and mitochondrial DNA sequences. Biochem Syst Ecol 25:81–97

BIBLIOGRAPHY

Tabor KL, Brewster CC and Fell RD (2004) Analysis of the successional patterns of insects on carrion in southwest Virginia. J Med Entomol 41(4):785–795

Tamura K, Dudley J, Nei M and Kumar S (2007) MEGA4: Molecular Evolutionary Genetics Analysis (MEGA) software version 4.0. Mol Biol Evol 24(8):1596–1599

Tangioshi L, Browne R, Hoyt S and Lagier R (1976) Empirical analysis of variable temperature regimes of life stage development and population growth of *Tetranychus mcdanieli* (Acarina: Tetranychidae). Ann Entomol Soc Am 69:712–716

Tao S (1927) A Comparative Study of the early Larval Stages of some common Flies. Am J Epidemiol 7:735 – 761

Tarone AM and Foran DR (2008) Generalized additive models and Lucilia sericata growth: assessing confidence intervals and error rates in forensic entomology. J Forensic Sci 53(4):942–948

Tomberlin JK and Adler PH (1998) Seasonal colonization and decomposition of rat carrion in water and on land in an open field in South Carolina. J Med Entomol 35(5):704–709

Tourle R, Downie DA and Villet MH (2009) Flies in the ointment: a morphological and molecular comparison of Lucilia cuprina and Lucilia sericata (Diptera: Calliphoridae) in South Africa. Med Vet Entomol 23(1):6–14

Trumble J and Pienkowski R (1979) Development and Survival of Megaselia scalaris (Diptera: Phoridae) at selected temperatures and photoperiods. Proc Entomol Soc Wash 81(2):207–210

Turner B and Wiltshire P (1999) Experimental validation of forensic evidence: a study of the decomposition of buried pigs in a heavy clay soil. Forensic Sci Int 101(2):113–122

VanLaerhoven SL and Anderson GS (1999) Insect succession on buried carrion in two biogeoclimatic zones of British Columbia. J Forensic Sci 44(1):32–43

Vincent S, Vian JM and Carlotti MP (2000) Partial sequencing of the cytochrome oxydase b subunit gene I: a tool for the identification of European species of blow flies for postmortem interval estimation. J Forensic Sci 45(4):820–823

Vinogradova EB and Kaufman BZ (1995) Photo- and thermopreferenda in the blowfly Calliphora vicina R.-D. (Diptera, Calliphoridae). Entomological Review 74 (7):16–24

Voss S, Forbes S and Dadour I (2008) Decomposition and insect succession on cadavers inside a vehicle environment. Forens Sci Med Pathol 4 (1):22–32

Wallman J and Adams M (1997) Molecular systematics of Australian carrion-breeding blowflies of the genus Calliphora (Diptera: Calliphoridae). Austr J Zool 45:337–356

Wallman JF and Donnellan SC (2001) The utility of mitochondrial DNA sequences for the identification of forensically important blowflies (Diptera: Calliphoridae) in southeastern Australia. Forensic Sci Int 120(1-2):60–67

Wallman JF, Leys R and Hagendoorn K (2005) Molecular systematics of Australien carrion-breeding blowflies (Diptera: Calliphoridae) based on mitochondrial DNA. Invertebrade Systematics 19:1–15

Watson EJ and Carlton CE (2005) Insect succession and decomposition of wildlife carcasses during fall and winter in Louisiana. J Med Entomol 42(2):193–203

Waugh J (2007) DNA barcoding in animal species: progress, potential and pitfalls. Bioessays 29(2):188–197

Wehner R and Gehring W (1995) Zoologie. Thieme

Weinland E (1906) Über die Stoffumsetzung während der Metarmophose der Fleischfliege (Calliphora vomitoria). Z Biol 47:186–231

Weissmann L and Podmanická D (1970) [Influence of nutrition and breeding temperature on the lenght of larval development of Scotia segetum Den. et Schiff]. Biologia (Bratisl) 25(8):537–545

Wells J and Williams D (2005) Validation of a DNA-based method for identifying Chrysomyinae (Diptera: Calliphoridae) used in a death investigation. Int J Legal Med :1–8

Wells JD and Sperling FA (2001) DNA-based identification of forensically important Chrysomyinae (Diptera: Calliphoridae). Forensic Sci Int 120(1-2):110–115

Wells JD and Stevens JR (2008) Application of DNA-Based Methods in Forensic Entomology. Annu Rev Entomol 53:103–120

Wells JD, Wall R and Stevens JR (2007) Phylogenetic analysis of forensically important Lucilia flies based on cytochrome oxidase I sequence: a cautionary tale for forensic species determination. Int J Legal Med Sci Law 121:229–233

Wetzel W, Reibe S and Madea B (2009) [An entomological case report during the winter months: estimation of the post-mortem interval considering the influence of cold temperatures on the development of the forensically important blowfly Calliphora vomitoria]. Arch Kriminol 223(3-4):123–130

Wooldridge J, Scrase L and Wall R (2007) Flight activity of the blowflies, Calliphora vomitoria and Lucilia sericata, in the dark. Forensic Sci Int 172(2-3):94–97

Worner S (1992) Performance of Phenological Models under variable Temperature Regimes: Consequences of the Kaufmann or Rate Summation Effect. Environmental Entomology 21(4):689–99

www.ingramcontent.com/pod-product-compliance
Lightning Source LLC
Chambersburg PA
CBHW021104210326
41598CB00016B/1318